Henry William Chisholm

On the Science of Weighing and Measuring and Standards of Measure and Weight

Henry William Chisholm

On the Science of Weighing and Measuring and Standards of Measure and Weight

ISBN/EAN: 9783337337940

Printed in Europe, USA, Canada, Australia, Japan

Cover: Foto ©berggeist007 / pixelio.de

More available books at **www.hansebooks.com**

Henry William Chisholm

On the Science of Weighing and Measuring and Standards of Measure and Weight

ISBN/EAN: 9783337337940

Printed in Europe, USA, Canada, Australia, Japan

Cover: Foto ©berggeist007 / pixelio.de

More available books at **www.hansebooks.com**

NATURE SERIES.

ON THE SCIENCE OF

WEIGHING AND MEASURING

AND

STANDARDS OF MEASURE AND WEIGHT.

BY

H. W. CHISHOLM,
WARDEN OF THE STANDARDS.

WITH NUMEROUS ILLUSTRATIONS.

London:
MACMILLAN AND CO.
1877.

[*The Right of Translation and Reproduction is Reserved.*]

CONTENTS.

GENERAL HEADS.

	PAGE
INTRODUCTION	1
I.—DEFINITION OF WEIGHT AND OF MEASURE .	4
II.—ANCIENT STANDARDS OF WEIGHT AND MEASURE	21
III.—ENGLISH STANDARD UNITS OF WEIGHTS AND MEASURES	48
IV.—THE RESTORED STANDARDS—IMPERIAL STANDARD POUND AND YARD	69
V.—SECONDARY IMPERIAL STANDARDS, INCLUDING MULTIPLES AND PARTS OF STANDARD UNITS	93
VI.—THE METRIC SYSTEM	97
VII.—WEIGHING AND MEASURING INSTRUMENTS, AND THEIR SCIENTIFIC USE	130
CONCLUSION	192

DETAILED LIST OF CONTENTS.

		PAGE
1.	Introduction	1
2.	General Heads of the subject	2

I. DEFINITION OF WEIGHT AND OF MEASURE.

3.	Weight and force of gravitation	4
4.	Effect of earth's ellipticity on the force of gravitation in different latitudes	4
5.	Intensity of gravitation, shown by seconds pendulum	6
6.	Length of seconds pendulum at London	6
7.	Local variations of intensity from differences in earth's density	8
8.	Latest determination of earth's ellipticity and dimensions from measurement of meridians	9
9.	Definition of measure.—Measure of extension	10
10.	Measure of volume.—Density of bodies	12
11.	Allowance in scientific weighings for weight of air displaced by bodies differing in density	14
12.	Illustrations of weight of air displaced by 1 lb. weights of various densities	16
13.	Determination of weight of normal measure of water at maximum density	17

II. ANCIENT STANDARDS OF WEIGHT AND MEASURE.

14.	Scientific use of accurate standards unknown in ancient times	21
15.	And ancient standards constructed without regard to scientific accuracy	23

x CONTENTS.

	PAGE
16. Authoritative works on ancient standards of weight and measure	23
17. Unfounded hypothesis of ancient Egyptian standards being based on earth's dimensions	24
18. Earliest standards of length based on proportions of human figure	27
19. The natural cubit the first standard unit of measure . .	28
20. The ancient Egyptian cubit and foot	28
21. Description of earliest known standard cubit, with divisions of foot and digits.	30
22. Hebrew standard measures of length	33
23. Babylonian or Chaldean system	34
24. Greek foot and Roman foot from which modern European foot derived	36
25. Ancient standard units of weight. The talent, mina, and shekel	36
26. Egyptian monetary and commercial standards of weight .	37
27. Hebrew system of weight derived from Egyptian . . .	38
28. Ancient Babylonian and Assyrian standards of weight. Lion and duck standard weights found at Nineveh . . .	40
29. Great Alexandrian talent and mina of the Ptolemies . .	41
30. Carthaginian or Bosphoric mina	42
31. Greek systems of weight. The Attic talent and mina . .	43
32. Roman weight. The libra or pound	44
33. Ancient standard units of measures of capacity . . .	44
34. Tabular summary of ancient and modern standard units of weights and measures	45

III. ENGLISH STANDARD UNITS OF WEIGHTS AND MEASURES.

35. Origin of English standard units	48
36. The yard, unit of length	50
37. The pound, unit of weight. Saxon money pound . .	54
38. Troy pound	56
39. The troy grain	56
40. Ancient merchant's pound, afterwards superseded by avoirdupois pound	57

CONTENTS.

xi

	PAGE
41. Exchequer standards of troy and avoirdupois weight	59
42. English standard measures of capacity	65
43. Imperial standard yard and pound, destroyed in 1834, and restored in 1854	67

IV. RESTORED STANDARDS OF THE IMPERIAL POUND AND YARD.

44. Proceedings of commission for restoring lost imperial standards	69
45. Construction of new imperial standard unit of weight	73
46. The lost standard pound and its authoritative copies	73
47. Weight of new standard derived from two platinum troy lbs. Sp and RS	75
48. New platinum troy lb. T, and auxiliary platinum weights	76
49. Mode of construction and verification of new imperial standard pound PS.	78
50. And of four platinum parliamentary copies, and thirty-six bronze gilt standard lbs.	79
51. Theoretical commercial standard lb. of brass, for regulating weighings in air	79
52. Construction of new imperial standard yard	81
53. The lost standard yard, and other authoritative standard yards	82
54. Length of new standard yard derived from authoritative standards	86
55. Mode of construction and verification of new imperial standard yard, and of parliamentary and other copies	87
56. Final report and recommendations of commission for restoring the standards	90
57. Provisions of Standards Act of 1855. Deposit of new standards at Exchequer	90
58. Creation of Standards Department in 1866; new scientific duties in relation to the standards	91
59. Sir J. Herschel's proposed geometrical inch, $\frac{1}{500,000,000}$ of polar axis	91

V. SECONDARY IMPERIAL STANDARDS, INCLUDING MULTIPLES AND PARTS OF STANDARD UNITS.

                                                          PAGE

60. Legal secondary standards of imperial weights and measures .   93
61. List of secondary standards now legalised 94

VI. THE METRIC SYSTEM.

62. Its scientific construction, and uniform decimal scale . . 97
63. Its origin, and proceedings taken for its establishment . . 98
64. Determination of standard unit of metric length, $\frac{1}{10,000,000}$ of meridian quadrant .   .   .   -   104
65. Construction and verification of the primary *Mètre des Archives*, and its principal copies 104
66. Standard unit of metric weight, based upon weight of cubic decimetre of water 106
67. Construction and verification of the primary *Kilogramme des Archives*, and its principal copies 112
68. The *Litre*, unit of metric capacity, and other derived units of superficial and solid metric measures 114
69. Number and denominations of metric weights and measures, having corresponding legal standards 115
70. English platinum standard metres of the Royal Society .   . 116
71. Kater's determination of imperial equivalent of the metre   . 117
72. Professor Miller's imperial equivalent of the kilogramme   . 119
73. Platinum and gilt gun-metal standard kilograms of the Standards department 121
74. Nomenclature of metric units, with their imperial equivalents 122
75. Progress of metric system, and of its adoption as an international system 124
76. Construction of new international metric standards by scientific commission at Paris 125
77. Metric convention of May 20, 1875, and establishment of international metric bureau at Paris 128

VII. WEIGHING AND MEASURING INSTRUMENTS, AND THEIR SCIENTIFIC USE.

	PAGE
78. The balance.—Two classes of balances; with equal, and with unequal arms	130
79. Ancient Roman balance, or steelyard; Egyptian equal-armed balance	132
80. Requirements in a just equal-armed balance	135
81. Balances of precision. Their construction on scientific principles	139
82. Illustration of kilogram balance of Standards department	141
83. Mendeleef balance, with short arms, and available as a vacuum balance	143
84. Relative advantages of a long beam and short beam in a balance of precision	143
85. Pointer and index scale of a balance of precision	146
86. Barrow's balance, used by Prof. Miller for all the weighings during the construction of the imperial standard lb.	147
87. Steinheil's method of reading the index scale reflected from a mirror on the beam of his balance	148
88. Two methods of scientific weighing: Borda's of substitution, and Gauss's of alteration	149
89. These methods described; superiority of Gauss's method proved	151
90. Illustration of comparison of standard weights by Gauss's method	155
91. Allowance for weight of air displaced by standard weights	155
92. Normal weight of litre of dry air, from Regnault's observations at Paris	156
93. Allowance for differences of latitude and of height above sea-level	156
94. And for differences of temperature, barometric pressure, and moisture	159
95. Actual mode of determining weight of air displaced, with illustration	160
96. Table of weight of air displaced by standard weights of various densities	162

	PAGE
97. Determination of volume or density of a standard weight by hydrostatic weighing	163
98. Illustrated by an example	167
99. Weighing standard weights in a vacuum	168
100. Deleuil's vacuum balance in use at Paris	169
101. New vacuum balance of Standards department	170
102. Determination of measure of capacity by weighing contents of water	174
103. Allowances for differences from normal temperature and barometric pressure	175
104. Determination of capacity by measurement of contents of water	177
105. Instruments for comparing standard measures of length: the beam compass	178
106. The micrometer microscope. Principles of construction	179
107. Single micrometer microscope, with apparatus for longitudinal displacement	181
108. Comparing apparatus with two micrometer microscopes, and transverse displacement	183
109. Description of comparing apparatus used for verifying the new imperial standard yard	183
110. Its application both to line-standard and end-standard yards.	185
111. New micrometrical comparing apparatus of Standards department	187
112. Lever frame with rollers for supporting standard bars	187
113. Computation of probable errors in results of comparisons of standards	190
114. Conclusion	192

LIST OF ILLUSTRATIONS.

FIG.		PAGE
1.	Imperial standard pound of platinum	16
2.	Official secondary standard pound of gilt gun-metal	16
3.	Official secondary standard pound of quartz	19
4.	Ancient Egyptian standard cubit of Amenemopht	31
5.	Transverse section of ditto	32
6.	Assyrian bronze lion standard weight	40
7.	Assyrian stone duck standard weight	40
8.	Exchequer standard yard of Henry VII. (left-hand end), with sixteenths of yard	51
9.	Exchequer standard yard of Henry VII. (right-hand end), with subdivisions of inches	51
10.	Exchequer standard yard rod of Queen Elizabeth (left-hand end) with sixteenths of yard	51
11.	Exchequer standard yard and ell bed of Queen Elizabeth (right-hand end) with inches of yard-bed	52
12.	Exchequer standard ell rod of Queen Elizabeth (left-hand end) with sixteenths of ell of 45 in.	52
13.	Exeter standard 14 lb. avoirdupois of Henry VII.	59
14.	Exchequer standard 112 lb. of Queen Elizabeth	60
15.	Exchequer standard troy lb. of Queen Elizabeth, formed of 8 oz. and 4 oz. weights	61
16.	Exchequer standard 1 lb. avoirdupois of Queen Elizabeth, bell-shape	62
17.	Exchequer standard 1 lb. avoirdupois of Queen Elizabeth, flat disc shape	63
18.	Form of modern local standard 1 lb. avoirdupois	64

LIST OF ILLUSTRATIONS.

FIG.		PAGE
19.	Exchequer standard Winchester bushel of Henry VII.	65
20.	Exchequer standard Winchester corn gallon of Henry VII.	66
21.	Exchequer standard quart of Queen Elizabeth	67
22.	Exchequer standard wine gallon of Queen Anne	68
23.	Standard troy lb. of 1758	69
24.	Platinum troy lb. of the Royal Society, RS	76
25.	Platinum troy lb. of the Standards department, T	76
26.	French standard toise of Peru	102
27.	Cylinder used for determining weight of cubic decimetre of water	108
28.	Platinum line-standard metre of the Royal Society	117
29.	Decimetre and nearly equivalent length of 4 imperial inches	119
30.	English platinum standard kilogram ℭ	121
31.	English gilt gun-metal standard kilogram ℜ	122
32.	Transverse section of new international line standard metre	126
33.	Transverse section of new international end standard metre	126
34.	Transverse section of mètre des archives	126
35.	Ancient standard Roman balance of Vespasian	133
36.	Ancient Egyptian equal-armed balance	134
37.	Kilogram balance of Standards department	142
38.	Mendeleef balance of Standards department, with short arms, and available as a vacuum balance	144
39.	Mode of hydrostatic weighing	164
40.	Deleuil's vacuum balance	169
41.	New vacuum balance of Standards department	173
42.	Field of micrometer microscope, of comparing apparatus for standards of length	181
43.	Mode of comparing end-standards with a line-standard measure	185
44.	Lever frame for support of standard bars (section)	188
45.	Ditto (plan)	189

ON THE

SCIENCE OF WEIGHING & MEASURING,

AND

STANDARDS OF MEASURE AND WEIGHT.

INTRODUCTION.

1. DURING the last few years public attention has been frequently drawn to the subject of our national weights and measures. The administrative and social questions of the improvement of our existing system of commercial weights and measures, and of the proposed introduction into this country of the decimal metric system—first established in France, and now being generally adopted on the continent of Europe, and, indeed, extending to the other quarters of the world—have formed the subjects of debate in successive sessions of Parliament, and are still awaiting solution. The scientific questions involved in the use of weights and measures have for a much longer period engaged the attention of many of our most eminent men of science, several of whom have been members of the various Standards Commissions from time to time appointed by the British Government.

These questions have also been the objects of investigation and deliberation by the large body of scientific men from all civilised countries, who were appointed members of the International Metric Commission at Paris. It may therefore be useful to bring together and place before the public the several points involved in the science of weighing and measuring, and to give some account of our standards of weight and measure, as well as to describe in some detail the scientific construction of our existing imperial standard yard and pound. No sufficient means have hitherto been adopted for making the general public acquainted with these parts of a subject in which they are so directly interested; the information hitherto published respecting it having been confined to a few papers in the *Philosophical Transactions*, Reports of the several Standards Commissions, and other Parliamentary Returns. Of these papers, the most important are the accounts of the construction of the imperial standard pound, by Prof. W. H. Miller, in *Philosophical Transactions*, 1856, and of the construction of the new imperial standard of length, by the Astronomer Royal, now Sir G. B. Airy, K.C.B., in 1857. In the following treatment of the subject, use will be made of these papers, as well as of other authoritative works relating to weights and measures.

2. The science of weighing and measuring comprehends the following points:—

The scientific definition of weight and measure.

The authoritative establishment of fundamental units of weight and measure of length, and the construction of their material representatives as primary standards, in relation to which all numerical amounts of weight and measure are to be expressed.

The establishment of determinate aliquot parts and multiples of the primary units of weight and measure, as well as of other units derived from them, such as the unit of measure of capacity, &c.; and the construction and verification of their material representatives, as secondary standards, by comparison with which the accuracy of all weights and measures in ordinary use is to be determined.

The scientific methods of using standard and other weights and measures in which special accuracy is required, as well as auxiliary scientific instruments, such as balances, thermometers, barometers, micrometers, and other comparing apparatus.

And, lastly, the determination of the just results of weighing and measuring with these scientific instruments, after allowing for all indirect influences affecting the accurate direct results of weighing and measuring. For instance, differences arising from the physical composition of bodies, variations of temperature and consequent expansion or contraction of the several substances, changes of condition in the medium in which the comparisons are made, &c., including also a computation of the probable errors of the final results.

The whole subject will be treated under the following general heads :—

I. Definition of weight and of measure.

II. Ancient standards of weight and measure.

III. English standard units of weights and measures.

IV. The restored standards, imperial standard pound and yard.

V. Secondary imperial standards, including multiples and parts of standard units.

VI. Derived units of imperial weight and measure, with scale of multiples and parts.

VII. The metric system.

VIII. Weighing and measuring instruments and their scientific use.

I. *Definition of Weight and of Measure.*

3. The *weight* of a body is the measure of the force of gravitation which the mass of our globe exercises upon the mass of all smaller bodies upon its surface, and in a line perpendicular to the surface of the earth, or, more strictly speaking, to the surface of a still liquid upon the earth's surface.

Gravitation is the effect of the force of attraction inherent in all physical bodies in our solar system, by which they are drawn towards each other in proportion to their *mass*, or the quantity of matter that each body contains. This attraction acts also with a force varying inversely as the square of the distance between the centres of each mass, and with a velocity in proportion as the medium through which the bodies are drawn is more or less rare.

4. If our globe were a perfect sphere and of uniform density, the force of gravitation would be the same on all parts of the earth's surface. But it is known that the figure of the earth is an oblate spheroid, or a sphere flattened at the poles. The latest and most

authoritative computations from actual geometric measurements have determined the amount of the flattening, or the difference of length of the polar axis, as compared with the mean equatorial diameter, at about the $\frac{1}{295}$th part of the earth's diameter. The effect of this difference of distance from the surface to the centre of the earth is to increase the force of gravitation in passing from the equator to the pole. It was shown by Newton that this increase of weight is nearly in proportion to the versed sine of double the latitude, in other words, to the square of the sine of the latitude. It has accordingly been computed that a body weighing 100 lbs. at the equator must weigh more than 100½ lbs. at either pole, and more than 100¼ lbs. in the latitude of London. The Astronomer Royal has, however, pointed out in his published *Lectures on Popular Astronomy* (Fifth Edit. p. 244), that the statement of a body weighing more at the pole than at the equator, though correct, is to be received with caution. If a pair of scales with proper weights were carried from the equator to the pole, the same weights which balanced a stone at the equator would also balance it at the pole; because the effect of the force of gravitation on both is altered in the same degree. But if a spring balance were carried from the equator to the pole, the spring would be more bent by the weight of the same stone at the pole than at the equator.

5. The weight of a body is also affected by the force of gravitation varying at places differing more or less in height above the mean level of the earth's surface, or the mean sea-level; and also at different points of

6   *WEIGHING AND MEASURING.*

the earth's surface, according to the density of the substance of the earth underneath these points.

In like manner, the weight of a body is affected by the density or specific gravity of the medium in which it is weighed. In accordance with the principle that all bodies immersed in a liquid suffer a loss of weight precisely equal to the weight of the liquid displaced, the discovery of which is ascribed to Archimedes, it was also demonstrated by Newton that a body immersed in any fluid specifically lighter than itself loses so much of its weight as is equal to the weight of a quantity of the fluid of the same bulk with itself. Hence a body loses more of its weight in a heavier fluid than in a lighter one, and therefore it weighs more in a lighter fluid than in a heavier one, for instance, more in air than in water.

The method of measuring the intensity of gravity on different parts of the terrestrial spheroid, by means of the seconds pendulum, is said to be originally due to Borda, as described in a Memoir inserted in vol. iii. of the *Base du Système Métrique.* From the results of Borda's experiments, made towards the close of the last century, Laplace computed the ellipticity of the earth to be $\frac{1}{336}$.

6. In the *Philosophical Transactions of the Royal Society* for 1818, Captain Kater stated the results of his pendulum experiments in London, and determined the length of the pendulum vibrating seconds, or completing one vibration in $\frac{1}{86400}$ part of a mean solar day, when measured in a vacuum at the mean level of the sea, and at a temperature of 62° Fahr., to be 39·13842 inches of the Standard yard, which was

legalised in 1824 as the Parliamentary Standard of length. The latitude of his place of observation in London was 51° 31′ 4″ N. He subsequently made a slight correction in this determination, making the length of the seconds pendulum to be 39·13929 inches, as shown in the *Philosophical Transactions*, 1819, and this length, or rather 39·1393 inches, was declared to be the true length in the Standards Act of 1824.

It was, however, discovered by Bessel that the correction which had ordinarily been applied, and was applied by Kater, for reducing the vibrations of a pendulum, as observed in ordinary air, to vibrations in a vacuum, ought to be greatly increased. The experiments were consequently repeated by Captain (afterwards Sir Edward) Sabine, with special reference to the form of pendulum usually employed in England. In the *Philosophical Transactions*, 1821, Sir Edward Sabine has shown, as the result of his experiments on the length of the seconds pendulum in Greenwich Observatory, that its length vibrating 86,400 seconds in the 24 hours, at the temperature of 62° Fahr., and in a vacuum, was found to be 39·13734 inches. Sir Edward Sabine has also shown, as the results of his experiments on the acceleration of the pendulum in different latitudes, that the mean diminution of the force of gravity from the pole to the equator was 0·0055138, in other words, that a weight of 100 lbs. at the equator would be increased by 0·55138 lb. at the pole; whilst the resulting mean ellipticity of the earth deduced from his pendulum observations, was $\frac{1}{313\cdot 6}$.

But the latest computations of other men of science

from further experiments with the seconds pendulum, make the ratio of increase of gravity from the equator to the pole to be 0·0051828, and the ellipticity of the earth about $\frac{1}{285}$.

7. In his paper on the Yard, the Pendulum, and the Metre, Sir J. Herschel has observed that the true measure of the earth's attraction (independent of centrifugal force arising from its rotation) is best to be derived from an ideal seconds pendulum supposed to vibrate at the extremity of the earth's polar axis; and that the mean length of the polar or of the equatorial pendulum must be derived from the general result of observations of the lines of oscillation of one and the same invariable pendulum at a multitude of geographical stations in all accessible latitudes in both hemispheres; but that no two combinations agree in giving the same precise length, in consequence of the local deviations of the intensity of gravity, due to the nature of the soil or crust of the earth, and the configuration of the ground immediately beneath and around the places of observation. And further, that since the pendulum cannot be observed at sea, the whole sea-covered surface of the globe is of necessity excluded from furnishing its quota of observations to the final or mean conclusion. Water being on the average not more than one-third the weight of an equal bulk of land, such as the earth's surface consists of, and only $\frac{2}{11}$ of the mean density of the globe, the force of gravity at the surface of the sea is less than at the sea-level on land by the attractive force of as much material taken at twice the specific gravity of water (or at $\frac{4}{11}$ that of the globe),

as would be required to raise the bottom to the surface. ⚹

8. With regard to the determination of the earth's ellipticity, as shown by actual measurements of the dimensions of our globe, and to the relative length of the equatorial diameter and the polar axis of the earth, the most recent determination is that by Colonel Clarke, as stated in his *Comparison of Standards of Length,* published in 1866. This memoir has been declared by Sir J. Herschel to be the most complete and comprehensive discussion yet received on the subject of the earth's figure, and to be held as the ultimatum of what scientific calculation is as yet enabled to exhibit as to its true dimensions and form.

Colonel Clarke's results were computed, not from pendulum experiments, but from the combination of all the separate measurements of arcs of meridians in Peru, France, Prussia, Russia, Cape of Good Hope, India, and in the United Kingdom. They are as follow :—

	Colonel Clarke's Computations.			Metres according to Capt. Kater's equivalent.
	Feet.	Inches.	Metres.	
Length of Polar axis	41,706,858	500,482,296	12,712,136	12,712,020
Longer equatorial axis (long. 15° 34' E.)	41,853,700	502,244,400	12,756,588	12,756,470
Shorter equatorial axis (long. 105° 34' E.)	41,839,958	502,079,496	12,752,701	12,752,588
Length of meridian quadrant of Paris	32,813,324	393,762,292	10,001,472	10,001,381
Length of minimum quadrant (long. 105° 34' E.)	32,808,772	393,704,064	10,000,024	9,999,953

In computing these equivalents, Colonel Clarke takes the metre at the temperature of 32° F. from his own measurements to be equal to 1·09362311 yard at 62°, that is to say to 3˙28086933 feet, or to 39˙37043196 inches, instead of the more generally received determination by Capt. Kater of 39˙37079 inches. The metric length according to both these equivalents is here given.

From this determination of the earth's dimensions, it was computed that the earth's ellipticity in the longitude of Paris was $\frac{1}{285}$, and its mean ellipticity in all longitudes $\frac{1}{298}$. This last computation has since been corrected, by allowing for local attraction, to $\frac{1}{295}$.

Hence also the mean length of a degree of latitude in the longitude of Paris is $\frac{32,813,524\cdot38}{90} = 364,591$ feet, or 69˙05 miles. The mean diameter of the earth is 41,800,173 feet, or 7216⅜ miles, and its mean circumference 23,871 miles.

Thus not only each longitudinal meridian, but also the equator is slightly elliptical.

Sir H. James states in his preface to Colonel Clarke's work, that the longest meridian in 15° 34' east longitude, nearly corresponds to the meridian in the eastern hemisphere which passes over the greatest quantity of land; and in the western hemisphere to that which passes over the greatest quantity of water, as it passes through the centre of the Pacific Ocean. The shortest meridian in 105° 34' east longitude nearly corresponds to that which passes over the greatest quantity of land in Asia; and in the western hemisphere, to that which passes over the greatest quantity of land in North and South America.

9. The connection here shown to exist between the definition of weight and the measurement of the

DEFINITION.

dimensions of our globe, leads naturally to the definition of measure.

Measure is generally understood to mean the determination of the dimensions of a body with relation to a fixed standard unit, or the measure of extension only ; and it is in this sense that it will now be taken in discussing the "science of measuring."

Strictly speaking, measure includes weight, which is the measure of the gravitation of bodies towards the centre of gravity. And measures of capacity also are almost universally derived, not from their cubical dimensions, but from the weight of pure water contained in them under determinate conditions as to temperature and atmospheric pressure.

The measure of temperature is based upon the observed rate of linear expansion by heat of a body selected for this purpose (generally mercury), taking as constant units the temperature of melting snow or ice, and of water boiling under determinate atmospheric pressure.

In defining measure, it should be added that it is also applied to the measure, or (as it is termed) *admeasurement* of the tonnage of ships, being a determination of the weight a ship is capable of carrying with relation to its measure of cubic capacity ; to value in relation to a monetary unit; to time and duration in relation to the unit of a mean solar day or of a second, its 86,400th part; to velocity, by combining the measure of extension with that of time or duration ; to mechanical work, the unit of which is a horse-power, as it is commonly termed, or more properly the power of raising 33,000 lbs. one foot in one minute, thus combining the measures of linear extension, weight, and time; to angles the unit being a degree of the 360th part of a circle described from the point of junction of the two straight converging lines forming the angle; &c., &c. It is not, however, proposed here to refer further to these measures, or to the scientific questions connected with them.

The measure of extension comprehends

The measure of length, or linear extension;

The measure of surface, or square measure;

The measure of volume, or solid or cubic measure;

The measure of capacity, or the cubical quantity contained in any vessel for measuring dry substances, liquids, or aëriform fluids.

All these measures of extension are based upon one fixed standard unit of length; and as all measures of length vary according to their temperature from expansion or contraction, the length of the standard must be fixed at a normal temperature.

10. The measure of volume or bulk of a body, as compared with that of another body differing in volume but equal in weight, is also expressed by its density, which in like manner is stated in relation to a fixed standard unit. The densities of bodies are in the direct ratios of their masses, or quantity of matter, and in the inverse ratios of their volume.

The density of a body is defined to be the mass contained in a unit of volume, when referred to a uniform standard. The density is to be distinguished from its specific gravity, which shows its weight in relation to its volume. The specific gravity of a body is defined to be the *weight* of a unit of its volume, also when referred to a uniform standard.

DEFINITION.

The specific gravity of a body is the quotient of its density when divided by the density of that substance which is considered as unity. Pure water is generally adopted as such unity. But the density of all bodies varies with their temperature, because the same invariable quantity of matter which the body contains is always distributed over its whole volume, and this is variable with the temperature; so that, generally speaking (with some exceptions, pure water, for instance, at certain temperatures), the body, at a higher temperature, has less density than at a lower temperature. We must therefore fix a certain temperature at which the body, as well as the water, must be considered with regard to their density. It is not necessary that this fixed temperature should be the same for the body and the water, its choice for both being quite arbitrary.

It is obvious that the most convenient standard temperature for expressing the density of bodies is that of one of the fixed points of the thermometer; and the temperature of melting ice or snow (32° F. or 0° C.) is generally adopted for all bodies except water. But for pure water there is a maximum of density which occurs at nearly 39° F. or 4° C., and this maximum density of pure water is taken as the unit of density.

The sign Δ prefixed to the symbol of any weight, with its numerical value following, denotes the ratio of the density of the weight at the temperature of melting snow to the maximum density of pure water.

The relation of the bulk or volume of a body to its weight is expressed both by its density and its specific gravity, these terms being often used indiscriminately. But the former term is more strictly applicable to solid bodies, and the latter to liquids and gases.

To ascertain the density of a body, it is requisite that its volume should be determined, as the density cannot be directly found. The actual volume may be determined—

(1.) Either by cubic measurement, when the form of the body admits of this measurement being actually made; this however occurs but rarely.

(2.) Or by ascertaining its specific gravity, from determining the difference of its weight when weighed in air and in water. This is the readiest and most accurate mode of determining both its volume and its density, but the immersion of a body in water is not always practicable, or may be injurious to the body under experiment.

11. It has already been mentioned that the gravitation or weight of bodies varies with their density and the density of the medium in which they are placed. In order to ascertain the true relative weight, as well as the actual weight of standard weights differing in density, when they are weighed in air, it is necessary to allow for the weight of air displaced by each. It thus becomes necessary to reduce these weighings to a vacuum, *by deducting from the apparent weight in air the weight of the volume of air displaced by each standard.*

Such a computation is based upon the ascertained weight of a given volume of dry air under certain conditions. But the weight of a given volume of air

DEFINITION.

is necessarily greater or less according to its temperature, the pressure of the atmosphere, and other conditions affecting it; and the following data are requisite for ascertaining the weight of air displaced by each standard.

(1.) The mean barometric pressure during the comparison reduced to 32° F. and corrected for the pressure of vapour and of carbonic acid gas in the air.

(2.) The mean temperature of the air during the weighings.

(3.) The density of the metal of which each standard weight is composed.

(4.) The co-efficients of expansion of the metals and of air.

(5.) The actual weight of each standard.

From data 1 and 2 the ratio of the density of the air to the maximum density of water must be ascertained. This ratio is also affected by the height in relation to the mean level of the sea, and the latitude of the place where the comparison is made, as the force of gravity differs accordingly. In practice the determination of the weight of air displaced by standard weights is easily and quickly effected, either by the more accurate mode of making the computations from the above-mentioned data, with the aid of a logarithmetical formula and tables for reduction of weighings, or approximately by special tables showing the mean weight of ordinary air displaced by standards of various densities. The mean ordinary air taken as the standard air in this country is of the normal temperature of 62° F., the barometer being at 30 inches, with the mercury reduced by computation to the temperature of 32° F.; the amount of aqueous vapour in the air being assumed to be two-thirds of the quantity in saturated air, and the amount of carbonic acid contained in it being taken at 0·0004 of its volume.

12. The actual mode of ascertaining the weight of air displaced by standard weights when compared by weighings in air will be described more at length under the head of Scientific Methods of Weighing. But some illustrations may here be given of the effect of the difference of density in standard weights upon their weight in ordinary air. The following figures of 1 lb. avoirdupois weights are of the actual form and size :—

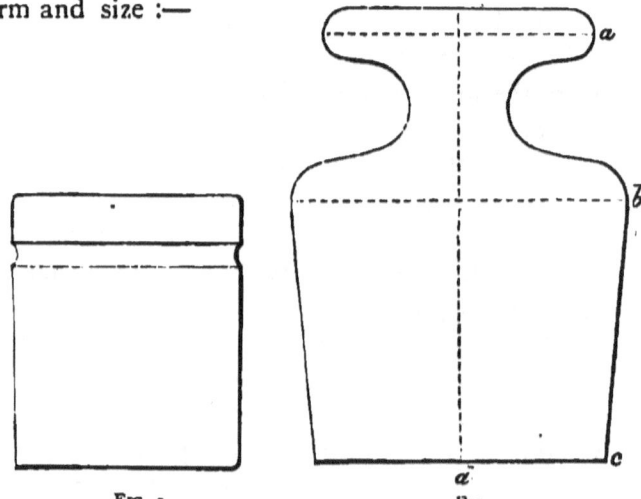

Fig. 1.
IMPERIAL STANDARD POUND OF PLATINUM. Cylindrical, with a groove.
Height = 1·35 inch.
Diameter = 1·15 inch.
Δ = 21·1572.
Displaces 0·403 grains of air.

Fig. 2.
OFFICIAL SECONDARY STANDARD POUND OF GILT GUN METAL. No. 31.
Diameter at a = 1·25 inch.
" " b = 1·65 "
" base c = 1·47 "
Height d = 2·2 inches.
Δ = 8·5144.
Displaces 1·001 grains of air.

It may here be seen that the difference of air displaced by the imperial standard lb. P.S. (Fig 1), and the gilt gun-metal lb. No. 31 (Fig. 2), is 0·598 grain ; and if they were equal in weight when in a vacuum, No. 31 would be 0·598 grain lighter in air of the given density. No. 31 is one of the gilt gun-metal secondary standard pound weights, intended to regulate the weighings in air of all commercial weights. As the primary platinum standard P.S. from its greater density displaced so much less air than ordinary brass and iron weights—the density of cast-iron being about 7·408, and a cast-iron pound displacing about 1·150 grain of air—the weight of all the gilt gun-metal pounds, of which No. 31 was one, was referred by Prof. Miller to a theoretical commercial standard pound of brass of the average density of brass and bronze weights ($\Delta = 8\cdot143$), and thus displacing 1·047 grain of standard air. This commercial standard pound, denoted as W., was assumed to be of the same weight in a vacuum as P.S., and consequently in standard air P.S. was 0·644 grain heavier than W. It has been adopted as the standard unit of all brass and bronze imperial weights used for weighing in ordinary air.

The standard pound of quartz (Fig. 3) displaces 3·217 grains of air. It was constructed as an auxiliary standard, on account of the invariability of quartz ; and its apparent weight in air was made intermediate between that of a pound of platinum and a pound of brass, being 0·447 grain lighter than P.S. and 0·197 grain heavier than W. in standard air.

13. As the determination of the density of bodies

has thus been referred to the maximum density of an equal volume of water, it was evidently necessary to determine the absolute weight of a normal measure of water at its maximum density, in order to determine the true weight in air of a given volume of any

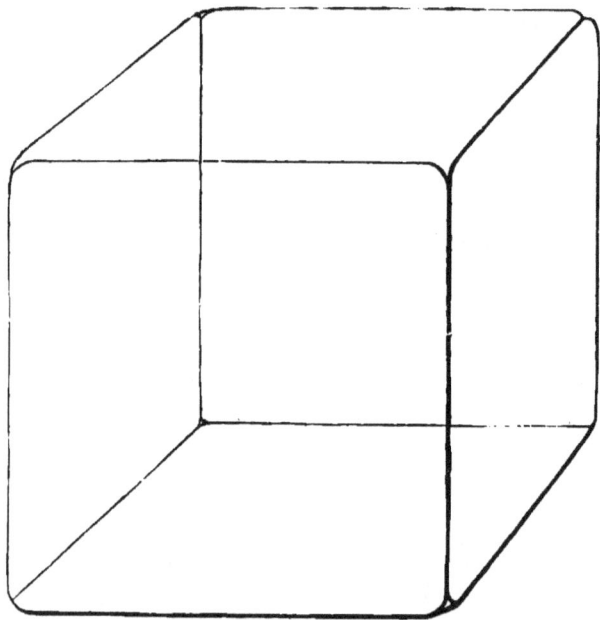

FIG. 3.—QUARTZ POUND IN STANDARDS DEPARTMENT.
Size = 2·17 inches cube, edges rounded.
Δ = 2·605. Displaces 3·216 grains of air.

substance from its ascertained density. It is claimed to be one of the important advantages of the decimal metric system, that this relation of weight to volume may be at once ascertained from the circumstance of the unit of weight, the kilogram, having been determined by its being the weight of a cubic decimetre of pure water at its maximum density. Thus the volume of any body expressed in cubic decimetres, or the measure of capacity of liquids expressed in litres (the litre being the measure of a vessel holding a cubic decimetre of water at its maximum density), when multiplied by its density, at once gives the weight in kilograms; or, if expressed in cubic centimetres, the weight will be given in grammes. For instance, a mass of cast-iron of the volume of two cubic decimetres, and with a density of 7·408, will weigh 14·816 kilograms; and half a litre (or 500 cubic centimetres) of mercury, the density of which is 13·59, will weigh 679·5 grammes.

There is not the same simple relation between the unit of weight and of volume or capacity in the imperial system as is found in the metric system. The relation between the unit of cubic capacity derived from the unit of length and the unit of weight has been determined experimentally in England from ascertaining the weight of a cubic inch of pure water; and the determination by Sir George Shuckburgh in 1798 was accepted by scientific men in this country, and has been legalised by statute, by which a cubic inch of water at the temperature of 62° F. weighed in air of the same temperature, with the barometer at 30 inches, is declared to weigh 252·458 grains of brass. From this ratio, the cubic capacity of the standard gallon, containing 10 lbs. weight of water, is declared to be 277·274 inches, and a cubic foot of water is declared to weigh 62·321 lbs. avoirdupois. But this ratio does not exactly agree with that adopted in France, nor indeed with other and different ratios

WEIGHING AND MEASURING.

adopted in Sweden, Austria, and Russia respectively, as determined from separate experiments made in each of these countries. For practical purposes, the cubic foot of water is frequently taken to be of the approximate weight of 1,000 avoirdupois ounces. Its legal weight in ordinary air is 997·136 oz., being only 0·2864 per cent. less than this assumed weight. As respects the metric system, even upon the assumption that the weight of a cubic decimetre of water is exactly a kilogram, according to its theoretical definition, as to which doubts exist, it is only equal to this weight when the water is at the temperature of about 39° F. or 4° C., and when weighed in a vacuum against brass weights at the temperature of melting ice. When the water is at the ordinary temperature (say 62° F.) and weighed against brass weights in ordinary air (say, the barometer at 30 inches), a cubic decimetre of water would weigh not a kilogram or 1,000 grammes, but about 998·717 grammes according to the French computation; the difference being the loss of weight caused by the greater weight of air displaced by a cubic decimetre of water than by its equipoise of brass weights. According to the English ratio, the cubic decimetre of water would weigh in air 998·680 grammes. And if the French ratio were applied to our imperial measures a cubic inch of water would weigh 252·336 grains, the capacity of the gallon would be 277·141 inches, and the cubic foot of water would weigh 62·291 lbs. But in point of fact, a new and authoritative international determination of the weight of a standard unit of water is very much needed, in order that its true weight may be satisfactorily ascertained and uniformly adopted in all countries.

II. *Ancient Standards of Weight and Measure.*

14. The use of weights and measures must have been one of the earliest necessities of civilised life. Josephus mentions the Jewish tradition that Cain, after his wanderings, built a city called Nod, and settled there, and was the author of weights and measures. The extreme antiquity of the use of weights and measures is also shown by the fact of the ancient heathens attributing the origin of weights and measures to the gods—the Egyptians to their god Theuth, or Thoth, and the Greeks to Mercury.'

But it is only within a comparatively recent period that weighing and measuring could justly be considered a scientific operation. It is to the general advancement of science, and more particularly to the voluntary and disinterested labours of scientific men, that the civilised world is indebted for improvements introduced from time to time in existing systems of weights and measures, as well as in methods of accurate weighing and measuring, and in the construction of instruments of precision. In this country very little indeed has been done by the government to obtain these advantages for the people, and for what has been actually accomplished the whole credit is due to the individual exertions of several of our most eminent men of science, and to the Royal Society as a scientific body.

The science of weighing and measuring, that is to say, the practical application of scientific methods to the construction and use of weights and measures, cannot be said to have existed before the last century, although the adaptation of the principle of the lever to the steelyard for weighing took place at an early period, this instrument being known as the Roman balance. A further and important step in the same direction was the discovery of the theory of specific gravity, and of its application to hydrostatic weighings, by Archimedes in the second century B.C. It was not, however, until after the discovery of the theory of universal gravitation by Sir Isaac Newton towards the close of the seventeenth century, and the discoveries earlier in the same century of the principle of the barometer by Galileo, and of the thermometer by the Accademia del Cimento at Florence, that means were first afforded of arriving at scientifically accurate results in weighing and measuring. We are indebted to Galileo's pupil, Torricelli, for the mercurial barometer and the discovery of atmospheric pressure, and to Sir Isaac Newton for fixing the determinate points of the thermometer at the temperatures of melting snow and boiling water. It was thus that Fahrenheit of Amsterdam was led to the construction of his mercurial thermometer, with its graduated scale, based upon the observed expansion of mercury from the lowest degree of cold observed in Iceland, and rising to the temperature of melting snow at $32°$, and up to that of boiling water at $212°$.

It is obvious that without a thermometer or any adequate means of determining the varying length of standard measures from dilatation or contraction by the effect of heat, and without a barometer or knowledge of the pressure and density of air, all measurings and weighings must have been wanting in scientific precision.

15. So far as can be ascertained from ancient records and from the evidence of such of the best authenticated standard weights and measures of earlier periods as are now known to be in existence, not only were the most authoritative standards constructed coarsely, and without regard to even ordinary precision, but no allowances were made, or thought of, for natural influences affecting their correct indication. No account appears to have been taken of the effects of variation of temperature in dilating or contracting measures of length; of the various densities of weights, and variable conditions of the air on the results of weighings; or of the different densities and conditions of water and other liquids used for determining the capacity of measures and the weight of their liquid contents. Nor indeed is there any record of instruments of precision having ever been employed for accurate comparisons of the old standard weights and measures. It is from such causes, and the necessary consequences, the want of uniformity found upon examining such specimens of the weights and measures of olden times as still exist, that so much doubt and uncertainty still prevail as to the absolute value of the ancient standard units of weight and measure in the different countries.

16. For information relating to ancient standards

of weights and measures, the following authoritative works have been chiefly relied on :—

Sir Isaac Newton's "Dissertation on Cubits." (Translated from the Latin, in vol. 2, p. 340, of Professor Piazzi Smyth's "Life and Work at the Great Pyramid," Edinburgh, 1867.

"Métrologie, ou Traité des Mésures, Poids, et Monnoies des Anciens Peuples et des Modernes," par A. J. P. Paucton. 4to. Paris, 1780.

"Memoire sur le Système Metrique des Anciens Egyptiens," par E. Jomard. Folio. Paris, 1817.

"Traité de Métrologie Ancienne et Moderne," par M. Saigey. 12mo. Paris, 1834.

"Metrologische Untersuchungen über Gewichte, Münzfüsse, und Masse des Alterthums," von A. Boeckh. 8vo. Berlin, 1838.

"Dictionnaire Universelle des Poids et Mésures, Anciens et Moderns," par H. Doursther. 8vo. Bruxelles, 1840.

"Essai sur les Systèmes Metriques et Monétaires des Anciens Peuples," par Don V. Vasquez Queipo. 3 Vols. 8vo. Paris, 1859.

"Das Münz-Mass-und-Gewichtswesen in Vorderaien bis auf Alexander den Grossen," von J. Brandis. 8vo. Berlin, 1866.

These two last-mentioned works are in every respect the most important, as they contain not only the most recent information upon the subject, but also references to preceding authors.

17. Frequent attempts have from time to time been made by scientific men and other persons who have inquired into the subject, to show that the

oldest system of weights and measures of which we have most knowledge, the ancient Egyptian system, was formed upon a scientific basis; that is to say, that the unit of length, from which the units of capacity and weight are admitted to have been derived, was taken from a natural constant, the ascertained length of a degree of the meridian, as was the case in establishing the metric system in France. Thus Paucton states that "the base of the Great Pyramid was made the principal standard, it being the 500th part of a degree of the meridian, previously measured for this purpose." But although several other authorities concur in this hypothesis, it will be found upon examination to have no valid foundation, and to rest only on a mere coincidence discovered in the time of the Ptolemies, nearly twenty centuries after the building of the Great Pyramid—the date of which is generally assigned to be about 2,200 B.C.

It will presently be shown that the most ancient unit of length in Egypt under the Pharaohs was the natural or common cubit of six palms, and equal to about 18·24 English inches; concurrently with which was the royal cubit of Memphis, of seven palms and equal to about 20·67 inches; and that the earliest Chaldæan cubits were of similar length. The length of the later Ptolemaic cubit, or cubit Belady, was about 21·87 inches, and was the 400th part of the great Egyptian stadium, containing 600 Ptolemaic or Phyleterian feet, and this was found to be the length of one of the sides of the Great Pyramid, as measured in the time of the Ptolemies.

It was not however until the second century B.C.

that the astronomer Ptolemy first determined the length of a degree of the meridian to be 500 of these Egyptian stadia. He arrived at this determination from correcting the results of computations which had been previously made. The first of these was by Eratosthenes, who lived 275 B.C., and had computed the length of a degree to be 700 stadia, from actual measurements of the distance between Alexandria and Syene, and between Syene and the Isle of Meroe. Another determination was by Posidonius, more than a century later, who had computed the degree to be equal to 666 stadia, from the measured distance between Alexandria and Rhodes. There is no evidence whatever of any previous determination of the length of a degree of the meridian, the computation of Eratosthenes being the earliest on record of the dimensions of the earth. In point of fact, the Ptolemaic measurement of a side of the base of the Great Pyramid (600 Phyleterian feet, equal to 690 English feet) was not the measure of the true base, which has now been proved to be 760 English feet, its equivalent being 750 of the more ancient Egyptian or Pharaonic feet. Ptolemy's computation of the length of a degree of the meridian, agreeing with 364,280 English feet, is however very near indeed to the truth. The mean length of a degree of the meridian has already been shown (8) to be 364,591 English feet, whence the length of the base of the Great Pyramid appears to be not $\frac{1}{500}$th, but about $\frac{1}{480}$th of a degree of the meridian. There is thus no real foundation for the hypothesis of the ancient unit of measure in Egypt having any such scientific basis as a natural constant derived from the dimensions of the earth. There can indeed be but little doubt that the earliest measures, the cubit and the foot, were taken from the proportions of the human body.

18. As regards the origin of standards of weight and measure, we learn from the most ancient records that the practice was to derive all other measures, as well as weights, from a recognised standard unit measure of length, the cube of which, or of a determinate aliquot part or multiple, formed the unit measure of capacity, and the weight of water or other liquid contained in this standard measure of capacity formed the unit of weight. We learn, too, not only from ancient records, but also from the very names of the measures of length, that the proportions of the human body were taken for indicating the several measures of length, and that the cubit, or length from the point of the elbow to the extremity of the middle finger, was practically adopted as the most convenient standard unit of length. The following scale of these natural proportions of the human body was most generally recognised, the digit, or breadth of the middle part of the first joint of the forefinger, being the lowest unit of the scale, viz. :—

The Digit	=	1 part
,, Palm, or Handbreadth	=	4 parts.
,, Span	=	12 ,,
,, Foot	=	16 ,,
,, Cubit	=	24 ,,
,, Step, or single pace	=	40 ,,
,, Double pace	=	80 ,,
,, Fathom, or length of extended arms from the tips of the fingers, nearly equal to the height of a man	=	96 ,,

19. The *cubit* is the only measure of length mentioned in the Book of Genesis as in use before the Flood. The earliest systems of weights and measures of which we now have any knowledge were the analogous systems established in Chaldæa, Egypt, and Phœnicia. All these systems were based upon the cubit as the standard unit measure of length, and it is to them that the derivation of the weights and measures used in almost all civilised countries can be traced. Of these early systems the Egyptian has probably had the greatest influence upon those of other countries. It is generally admitted that the Egyptian weights and measures passed into Asia and Judæa, as well as into Greece, and with some modifications extended to Italy, where they were adopted by the Romans, and subsequently by all European nations.

20. At the earliest recorded period of Egyptian history, two different cubits appear to have been in use, as computed from the external and internal dimensions of the Great Pyramid. The earliest of these two measures of length was the common, or natural cubit of six palms and twenty-four digits. Two-thirds of this cubit formed the ancient Egyptian foot. We have the evidence of the most ancient authors that the length of one of the sides of the square base of the Great Pyramid was 500 cubits, or 750 Egyptian feet. There is also the evidence of Sir H. James, the head of our Ordnance Survey Department, that the mean length of the side of the original base of the Great Pyramid is 760 English feet, according to the most authoritative measurements, the latest being by the Ordnance Surveyors in

1868. The length of the Egyptian foot is consequently shown to be equal to 1·013 English foot, or 12·16 inches (0·3086 metre), and the cubit to 18·24 English inches, or 0·463 metre. This cubit was identical with the Phœnician or Olympic cubit, afterwards adopted in Greece. The length of the ancient Greek cubit has been satisfactorily ascertained from a recent measurement of the Hecatompedon at Athens, the platform upon which the Parthenon stood. The Greek foot has thus been proved to be also equal to 12·16 English inches, and the Greek cubit, being one-half more, to 18·24 inches.

The second of the two ancient Egyptian cubits was the royal cubit, or cubit of Memphis, of seven palms or twenty-eight digits. Its length was computed by Sir Isaac Newton from Mr. Greaves's measurements of the internal dimensions of the Great Pyramid to be equal to 20·7 English inches. But since Newton's time other and conclusive evidence of the length of these two ancient Egyptian cubits has been brought to light. In addition to the well-known ancient Nilometer cubits, that of Elephantine, measuring according to Doursther nearly 20·7 English inches, there are no less than ten ancient standard cubit measures of the time of the Pharaohs still in existence. Some of these are made of wood, and in excellent preservation, others of stone more or less fractured, but still available as evidence of the ancient Egyptian measure of length. The antiquity and authenticity of these cubit measures are undoubted, the date of their construction extending back to a period not indeed of the building of the Great Pyramid, but yet more than 3,500 years ago. The mean length of the Egyptian royal cubit is thus ascertained to be 20·67 English inches, or 525 mm.

21. The most perfect of these ancient standard measures is the cubit of Amenemopht, an ancient cubit of hard wood, discovered amongst the ruins of Memphis in the early part of this century by M. Drovetti, Consul-General of France in Egypt. This ancient cubit is now deposited in the Museum of Turin. It bears the date of the reign of Horus, who is believed to have become King of Egypt about 1657 B.C., and to have been the ninth Pharaoh of the eighteenth dynasty. The section of this cubit measure is that of a rectangle with one edge bevelled off, as shown in Fig. 5. The measure itself is an end-standard royal cubit of seven palms or twenty-eight digits, with smaller subdivision lines. The natural cubit of six palms, or twenty-four digits, and its foot of sixteen digits, are also marked. (See Fig. 4.) The upper row of hieroglyphics on the lower part of the face, A, B, represents the different divinities to whom the several digits were consecrated. The second row, B, C (commencing on the left), gives the names in hieroglyphics of the several subdivisions of the cubit, viz.: the first and second palm divided into digits, the lesser and greater span, the foot, the natural cubit, and the royal cubit. The third row (commencing on the right) enumerates the digits, the fractional parts of which are shown in the fourth row, the first digit being divided into halves; the sixteenth digit is marked as completing the foot, or as being two-thirds of the natural cubit. Don V. Queipo, from whose work this

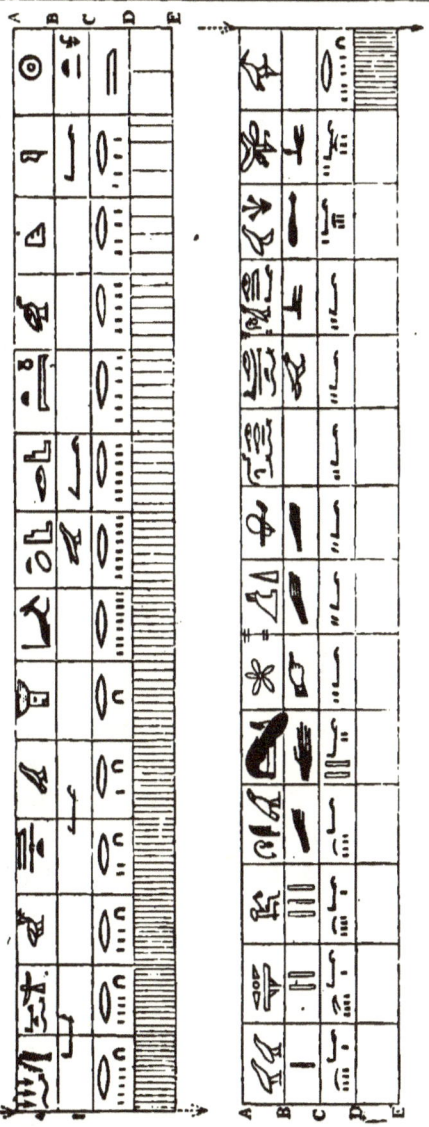

Fig. 4.—Ancient Egyptian Standard Cubit of Amenemopht. Half size.
(The cubit is in one length, and is here divided as indicated by -->.)

description of the cubit is taken, considers that the evidence thus adduced, confirmed as it is by that of the other ancient standard cubits, shows conclusively that the original Egyptian cubit was the common or natural cubit of twenty-four digits, whilst the sacred, or royal Egyptian cubit had an additional palm, or hand's-breadth, of four digits. Both Queipo and other authorities, such as Jomard, Boeckh, &c., notice the fact that the length of the additional palm of the royal Egyptian cubit is equal to about 2·5 English inches only, and is less than the average length of the

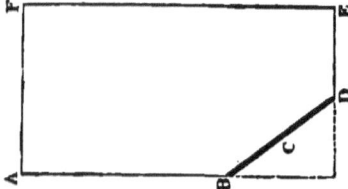

FIG. 5.—TRANSVERSE SECTION OF CUBIT. Full size.

six palms of the natural cubit, which slightly exceeds three of our inches. This discordance is undeniable. The relative proportion between the two cubits is not as 24 to 28, the number of digits contained in them respectively, but as 24 to 27·473. The probable explanation is, that the length of the Egyptian royal cubit was not taken directly from the common cubit, but from the ancient Chaldæan or Babylonian cubit, with which it was identical in length. In a paper read before the Royal Society on 19 June, 1873 (Phil. Trans. 1873), Sir H. James showed that the length both of the common cubit and the royal cubit of the ancient Egyptians was as above stated.

22. Sir Isaac Newton has stated his opinion that the Hebrews continued to use the measures of length brought with them from Egypt; and the evidence of Ezekiel (xliii. 14), about 574 B.C., shows that, like the royal cubit of Memphis, the cubit of the sanctuary consisted of "a cubit and a hand-breadth." There is much conflict of opinion as to the actual length of the several cubits in use by the Jews at different periods; but the fact that Moses always mentions the Egyptian measures of the cubit, span, and palm, as well as the Egyptian weights, the talent, with its smaller unit, the shekel, proves that the Hebrews originally brought their weights and measures from Egypt. According to Queipo four different cubits were used by the Jews. Besides the royal cubit of Memphis and the natural cubit of a man (Deut. iii. 11), used in the earlier periods, there was also the later Rabbinical cubit, estimated as equal to 21·85 inches (0·555 m.), and substantially the same as the Ptolemaic cubit Belady; and the cubit of the sanctuary mentioned by Ezekiel, which Queipo considers to have been one palm *more than the Rabbinical cubit*, and that this sacred cubit was equal to about 25·5 inches (0·648 m.). This was nearly the length of the Chaldæan and Persian cubit of eight palms, and the 10,000th part of the parasang, their unit of measure of distance, and nearly equal to four English miles, or 6·4 kilometres. The cubit mentioned by the Talmudists as the cubit of vessels, and used for the measurement of vessels of capacity, was the natural cubit, or cubit of a man, corresponding with the earliest Egyptian cubit.

In his dissertation on cubits, Sir Isaac Newton

states grounds for his opinion, that the sacred cubit of the Jews was equal to 24·7 of our inches, and that the royal cubit of Memphis was equivalent to five-sixths of this sacred Jewish cubit, or 20·6 inches.

If we take the natural cubit, or cubit of a man, mentioned in Deuteronomy, to be equal to 18·24 of our inches, the size of the iron bed of the giant Og, King of Bashan, stated to be nine cubits long and four cubits broad, must have been $13\frac{1}{2}$ feet by six. According to the reckoning of Maimonides, that a bed was usually one-third longer than the height of a man, Og must have been nine feet high; and the height of the giant Goliath of Gath, stated in the First Book of Samuel to be six cubits and a span, must have been nine feet four inches.

23. Dr. Brandis, who has investigated very closely the subject of the earliest standards of weight and measure, is of opinion that the Babylonian or Chaldæan system of weights and measures was the original system from which the Egyptian as well as that of other ancient nations was derived. He calls attention to the fact of the ancient Chaldæans having used not only the decimal system of notation, which is evidently the primitive system, derived from the use of the ten digits, but also a duodecimal system of reckoning, as shown by the division of the year into twelve months, the equinoctial day and night, each into twelve hours, the zodiac into twelve signs, &c. The duodecimal was combined with a sexagesimal system, by which the hour was divided into sixty minutes, the signs of the zodiac into thirty parts or degrees, and the corresponding division of the circle

into 360 degrees, with further sexagesimal subdivisions. He traces the origin of these two systems of reckoning to the observations of the heavenly bodies by the Chaldæan astronomers, which also led to the formation of the earliest system of weights and measures, founded upon their measure of time.

Besides the Chaldæan and Persian cubit of eight palms, referred to above, there was a more ancient Chaldæan or Babylonian cubit, mentioned by Herodotus as being three digits longer than the common cubit. Boeckh and Queipo, after quoting the ancient writers upon this point, concur in opinion that the three digits were Babylonian digits, which would make the Babylonian cubit equal to 20·4 English inches, or 0·527 metre, and substantially the same as the Egyptian royal cubit of Memphis. Numerous measurements of squared slabs of stone used in buildings at Babylon agree with a mean unit of measure of 20·67 inches, or 0·525 metre. The Babylonian foot is generally admitted to have been about $\frac{3}{5}$ of the Babylonian cubit, and equal to 12·6 inches, or 0·320 metre. This has been found to have been the unit of measure for Babylonian bricks, which are of uniform size. Queipo and Brandis concur in opinion that this Babylonian cubit was introduced into Egypt and adopted by the Egyptians as their royal cubit, together with the original standard measure of length, the natural cubit. This measure of the natural cubit may fairly be assumed to have been in use before the Deluge, and to have come down to the Chaldæans through Noah, who employed it as the unit of measure in the construction of the ark.

D 2

24. The Greek foot, equal to $\frac{2}{3}$ of the earliest Egyptian cubit, and to 12·16 English inches, or 0·308 metre, as already mentioned (20), was introduced from Greece into Italy, and was there divided into twelve parts or *unciæ*, according to the Roman duodecimal system, by which each unit of measure or weight, termed *as*, was divided into twelve *unciæ*. Our English words inch and ounce are thus derived from the Latin *uncia*. A modification of the length of the Greek foot was made by the Romans at a later period, and in the time of Pliny it was $\frac{1}{25}$th less than the more ancient foot, twenty-five Roman feet being equivalent to twenty-four Greek feet. Several ancient standards of this later Roman foot are still in existence, the mean length being equal to 11·65 English inches, or 296 millimetres. The modern measure of the foot in the different countries of Europe, with its duodecimal division of twelve inches, has been generally derived from the Greek foot and the Roman foot, some modifications having been made in its length in different localities and at different periods. The French foot, or *pied du roi*, is traditionally said to have been the length of Charlemagne's foot, as the English yard has been said to have been the length of Henry the First's arm. The *pied du roi* continued in use in France until superseded by the metre. It was equal to 12·789 English inches, or 0·325 metre.

25. No very satisfactory information exists as to the earliest system of weight in Chaldæa and in Egypt. The best authorities agree that, both in these countries and in other early civilised countries, the weight

ANCIENT STANDARDS.

of water or other liquid contained in the measure of the cubic foot constituted the larger standard unit of weight, the *talent*, and that a determinate aliquot part, the fiftieth, sixtieth, and in later periods the hundredth part, constituted the lesser unit of commercial weight, the *mina*. A third and smaller unit of weight for monetary purposes, the *shekel*, was derived from a similar aliquot part of the mina.

As the modern standard unit of weight, the pound, was derived from and is identical with the ancient mina, it may be interesting to examine more closely into the weight of this ancient standard unit. It is however to be observed, that the weight of the ancient mina varied not only in the several countries where it was commonly used, but also in the same country at different periods and even in different provinces, just as the modern European pound has varied during the period which has elapsed since the reign of Charlemagne. At the end of his third volume, Queipo gives a list of more than two hundred different European pounds. There was also the same practice in ancient countries of giving a different value to the same nominal unit of weight, according as it was used for monetary or commercial purposes, as has existed in this country up to the present time in the pound troy and the pound avoirdupois.

26. Assuming the most ancient Egyptian foot, two-thirds of the natural cubit, to have been equal to 1·013 English foot, or 308·6 mm., Queipo comes to the conclusion that the common Egyptian talent of the market, or *kikkar*, as it was termed, was equal to 29·36 kilos., or 64·728 lbs. avoirdupois; and the

commercial mina, its fiftieth part, equal to 587 grm., or 9,062 English grains.

The fact of the mina being in use as the ancient unit of weight in Egypt, and also as the unit of measure of capacity by the weight of liquid contents of vessels, is shown from an ancient inscription on the walls of the great temple at Karnac, recording the victories of Thothmes III., who, according to Sir Gardiner Wilkinson, reigned 1445 B.C. This inscription is said by Dr. Birch, in his Annals of this king, to record the number of mina weights of large quantities of wine, honey, spices, dates, and bitumen taken as booty in war or imposed as tribute.

Queipo is of opinion that the talent, the larger unit of Egyptian weight for monetary purposes, and for weighing the precious metals, was equal to the weight of water contained in the cube of $\frac{2}{3}$ of the royal or sacred cubit, and thus equivalent to 42·48 kilos. or 113·814 lbs. troy. He considers this to have been the weight of the Mosaic talent taken by the Hebrews out of Egypt. It was divided into fifty minas, each equal to 849·6 grm., or 13,111 English grains; and the mina into fifty shekels, each equal to 14·16 grm., or 218·5 English grains.

27. Some further knowledge of the ancient Egyptian system of weight is to be derived from the recorded weights of the Israelites, more particularly their monetary weights, which they must have brought with them at their exodus in 1491 B.C. according to the biblical chronology. At that period, and indeed in earlier times, we learn from the Mosaic record that in Egypt and other Eastern countries

ANCIENT STANDARDS.

the shekel was the unit of monetary weight. Payments were made in shekel weights of silver by Abraham. But the shekel was used not only as the unit for weighing silver (Gen. xxiii. 16, and Exod. xxx. 13) and gold (Gen. xxiv. 22), but also brass (Exod. xxxviii. 29), iron (1 Sam. xvii. 7), spices (Exod. xxx. 23, 24), and meat (Ezek. iv. 10). Queipo considers the values of the series of the earliest Hebrew weights, of which the shekel of the sanctuary was the unit, to have been as follows :—

		Shekel.	Imp. grains.		Gram.
(Ex. xxx. 13) ...	... Gerah	1/20	=	11 or	0·708
(1 Sam. ix. 8) ...	... Rebah	1/4	=	54·6 ,,	3·540
(Ex. xxxviii. 26)	... Bekah	1/2	=	109·25 ,,	7·080
	Shekel	1	=	218·5 ,,	14·160
(Ezek. xlv. 12)	... Mina	60	=	13,111 ,,	849·600
(Ex. xxxviii. 25, 26) Kikkar or Talent, 3,000, or 50 Mina = {113·814 lbs. troy} or 42480·000.

There appears to be satisfactory evidence from existing specimens of the earliest Jewish coins that the normal weight of the later Jewish shekel of silver was 218·5 troy grains, or 14·16 grammes. Thus the two talents of silver given by Naaman to Gehazi (about 894 B.C.), which were bound up in two bags with two changes of raiment, and laid upon two of Naaman's servants, must, if they were Israelitish talents, have each of them weighed 93·6 lbs. avoirdupois.

At a later period the Jews introduced a second system of weights, also derived from the shekel, but based upon the *drachma*, its fourth part, which was used in Greece and by neighbouring nations. From this new unit they formed a new mina of 100 drachmas, known as the common, or vulgar mina, together with a talent composed of sixty of these mina, the series of weights being as follows :—

	Drachma.		Grains.		Gram.
Pondiuscule	1/12	=	4·55	,,	0·295
Meʒh — ...	1/6	=	9·1	,,	0·59
Drachma, or Zuza ...	1	=	54·6	,,	3·54
Shekel	4	=	218·5	,,	14·16
Mina	100	=	5,463·0	,,	354·0
Talent	6,000	=	{56,907 lbs. troy}	or	21,240·0.

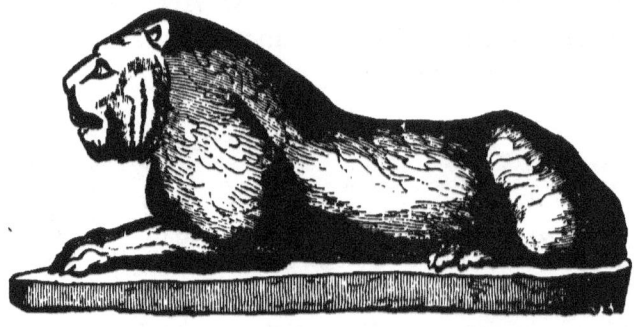

FIG. 6.—ASSYRIAN BRONZE LION STANDARD WEIGHT.

28. We have the most certain knowledge of the Babylonian and Assyrian system of commercial

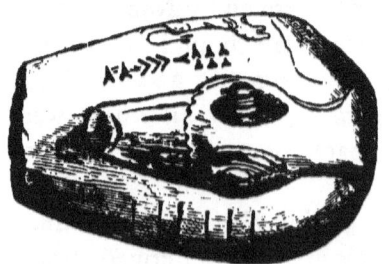

FIG. 7.—ASSYRIAN STONE DUCK STANDARD WEIGHT.

weight from the complete series of ancient standard weights discovered by Mr. Layard in the ruins of

Nineveh, and now in the British Museum. These constitute the earliest series of standard weights now known. There are two distinct series of weights, all bearing authoritative marks of their denominations and periods of construction. The principal series is of bronze, in the form of a crouching lion. The others are of marble or stone, of an oval or rounded form, representing a duck sitting with the head turned, and flattened on the back.

A full description of these ancient weights may be found in Layard's *Nineveh and Babylon* (vol. 1, p. 601), published by Mr. Murray, who has allowed the use of these illustrative figures.

These Nineveh weights represent two distinct systems, one exactly double the other. The mean weight of the mina of the larger series is 15,330 English gr. or 993·4 grm., the talent of sixty minas being equal to 59·604 kilogrammes.

29. In the reform of Egyptian weights and measures by Ptolemy Lagos there is some evidence that the former Egyptian system of weight was continued. The great Alexandrian talent of copper or brass is known to have been of the weight assigned by Queipo to the old Egyptian and Mosaic monetary talent, viz., 113·8 lbs. troy, or 42·480 kilos. It was divided into sixty minas, each = 10,926 gr. or 708 grm. And the mina into fifty shekels or didrachmas, each = 218 gr. or 14·16 grm. There was a smaller talent of silver half the weight of the great talent, which was divided into sixty minas, each = 5463 gr. or 354 grm., and the mina into 100 drachmas, each = 54·6 gr. or 3·54 grm. There are

now in the Louvre two ancient Egyptian weights of roughly rounded stones, bearing inscribed hieroglyphics, and weighing respectively 352.16 and 176.75 grm., evidently mina and half-mina weights. There are also three ancient Egyptian bronze weights, weighing respectively 3.57, 3.56, and 3.62 grm., evidently drachma weights.

The great Alexandrian talent was divided at a later period by the Romans, when they were in possession of Egypt, into 125 lbs. The Romans also divided the mina of the lesser Alexandrian talent into twelve ounces, each = 455 gr. or 29.5 grm., and several of these Egyptian ounce-weights are now at the Louvre. It is generally admitted that the diamond weight, the karat, of 3.1683 gr., or 0.2055 grm., was originally formed from being $\frac{1}{144}$ part of this Alexandrian ounce; and the fact is distinctly stated in a Syrian treatise on weights by Anania de Schiraz, written in the sixth century. The fourth part of this karat was the Alexandrian grain, equal to 0.79 troy grain.

30. Another system of monetary weight, established at Carthage, as well as in Macedonia and parts of Asia Minor, was known as the Bosphoric system, and was similar to, but distinct from, the Ptolemaic system. Under this system, the Olympic talent, equal to 78.662 troy lbs. or 29.360 kilos., was divided into 2,000 shekels, instead of 3,000, each shekel or tetradrachma being of the weight of 226.6 gr., or 14.685 grm., and the drachma one-fourth part, being equal to 56.6 gr. or 3.673 grm. There are numerous specimens existing of drachma and tetradrachma

silver coins of this system. The Bosphoric mina of one hundred drachmas weighed 5,666 gr., or 367 grm., and that has continued to be the weight of the mina or *yousdrouman* pound of Constantinople, up to the present time. It was this mina, formed into twelve ounces, that was sent by the Caliph Haroun al Raschid to Charlemagne, and upon it the French system of weight, *poids de marc*, was based.

31. The Attic mina was also used in Egypt to a great extent in the time of the Ptolemies. Out of the many and various systems of weight used in the different Grecian states, that based on the Attic talent, or Euboic, as it was also called, was the most in use, and is more frequently referred to by ancient authors. There is evidence that the earliest commercial Attic talent was derived, like the early Egyptian, from the fiftieth part of the weight of the Olympic cubic foot of water. It was equal to 29·325 kilos., or 64·65 lbs. avoirdupois. The Attic mina, its $\frac{1}{50}$th part, was equal to 9,051 troy gr., or 586·5 grm., and was divided into a hundred drachmas. But the best known Greek standard unit of weight was the Attic monetary mina, which was nearly the hundredth part of the great talent of Alexandria, or the earlier Pharaonic talent, and equal to 6,558 troy gr., or 425 grm. It was also divided into one hundred drachmas. Sixty of these Attic minas constituted the Attic monetary talent, of about 68·320 troy pounds, or 25·5 kilogs. The establishment of this system of monetary weight at Athens,

and the division of the mina into one hundred drachmas, were effected by Solon about 594 B.C., in order to relieve the people when oppressed by their debts to usurers, by lowering the weight of the money payable by them.

Another system of weight which was also current about the same period in Egypt was the Greek-Asiatic then in use in Persia, and throughout Asia Minor. The talent of this system was equal to 87·131 troy pounds, or 32·5 kilos., and was divided into one hundred minas, each of 5,015 gr. or 325 grm., and the mina into one hundred drachmas. During the Roman dominion in Egypt this mina appears to have been divided into twelve ounces, and there are now in the Louvre several of these old Egyptian ounce-weights.

32. The ancient Roman unit of weight was the libra, or pondus, from which the modern names of the livre and pound are derived. Its weight was equal to 5,015 Troy gr. or 325 grm., and it was identical with the Greek-Asiatic mina. The Roman larger unit of weight, the centipondium or hundred-weight, was of the same weight as the Greek-Asiatic talent. The Roman pound was also equal to the tenth part of the weight of water contained in the congius, the Roman unit of liquid capacity, and the cube of half the Roman foot.

33. As to the ancient units of measures of capacity, the difficulty of determining absolutely the values of both the primary and derived units, amongst the different opinions expressed by various authorities,

and with the little reliable information that has come down, is not less than the determination of the ancient units of weight.

The best evidence of the relative values of the units of capacity is to be found in the Mosaic writings, from which we learn that the *ephah*, or *bath*, was the unit of measures of capacity for both liquids and grain. The ephah is considered by Queipo to have been the measure of water contained in the ancient Egyptian cubic foot, and thus equivalent to 29·376 litres, or 6·468 imperial gallons, and to have been nearly identical with the ancient Egyptian *artaba*, and the Greek *metretes*. For liquids, the ephah was divided into six *hin*, and the twelfth part of the hin was the *log*. As a grain measure, the ephah was divided into ten *omers*, or *gomers*. The omer measure of manna gathered by the Israelites in the desert as a day's food for each adult person was thus equal to 2·6 imperial quarts. The largest measure of capacity both for liquids and dry commodities was the *cor* of twelve ephahs.

34. The limits of this work do not allow of a more detailed description of the ancient standards, nor of the standard units of weights and measures in foreign countries in more modern times. It may be sufficient to give the following tabular summary of the more important ancient and modern standard units of means of length, weight, and capacity.

STANDARD UNIT MEASURES OF LENGTH.

			Equivalents in Imperial Inches.
Egypt, Chaldæa—			
Natural Cubit	... ...	of 6 Palms (Foot ⅔ = 12·16 in.)	18·24 × 2 = 36·48
Royal Cubit...	... ...	of 7 Palms (Babylonian foot ⅔ = 12·16)	20·67 × 2 = 41·34
Chaldæa, Persia, and Later Arab—			
Hachemic Cubit	... ...	of 8 Palms	25·20
Judæa—			
Cubit of the Sanctuary	... ...	„ 8 Palms	25·50
Egypt under the Ptolemies—			
Cubit Belady	... ...	„ 7 Palms (Philetrian foot ⅔ = 14·6)	21·85
India—			
Ancient Hindoo Cubit	... ...	of 2 Spans	18·00
Judæa under the Romans—			
Talmud or Rabbinical Cubit	...	„ 7 Palms	21·85
Egypt under the Caliphs—			
Black Cubit ...	... ...	„ 7 Palms	21·34
Modern Egyptian Cubits—			
Pyk Charlie	... ...	„ 6 Palms	19·42
Pyk Balady	... ...	„ 7 Palms	22·94
Pyk Hindasah	... ...	„ 8 Palms	25·83
Pyk of Architects	... ...	„ 9 Palms	29·92
Ancient Rome—			
Pace (Mile = 1,000 Paces)	...	„ 5 Ft.(foot=11·65)	58·26
Prussia—			
Rhine Foot	... ...	„ 12 Zoll or Inches	12·36
China—			
Chid or Foot	... ...	„ 10 Parts	14·10
Russia—			
Archine (¼ of Sagene = 7 Eng. feet)		„ 16 Verschok	28·00
Roman States ...	... ...		
Braccio or Passetto	... ...	„ 3 Roman Palms	26·40
Austria—			
Klafter	... ...	„ 6 Austrian Feet	74·66
France—			
Toise ...	... ...	„ 6 Pieds du Roi	76·74
Metre	... ...	„ 10 Decimetres	39·37
Great Britain—			
Imperial Yard	... ...	„ 3 Feet	36·00 = Impl. Yard

STANDARD UNITS OF WEIGHT.

	Equivalents in Imperial	TALENT Grains.	MINA or POUND. Grains.	SHEKEL or TETRADRACHM. Grains.	DRACHM. Grains.
Chaldæan or Babylonian Talent of Silver	Royal	504,638	1/60 8,410	1/50 160	1/60 M. 84
Babylonian, Assyrian, and Phœnician	...	919,830	1/60 15,330	1/60 266	—
Ditto	Commercial	459,915	1/60 7,665	1/60 133	—
Babylonian	Gold	757,882	1/50 15,158	1/60 253	—
Egyptian, Hebrew, and Olympic	Monetary	655,566	1/50 13,111	1/60 218.5	—
Ditto	Commercial	453,093	1/50 9,062	1/60 151	—
Great Alexandrian Talent of Brass	Ditto	655,566	1/60 10,926	1/60 182	—
Lesser Alexandrian Talent of Silver (also later Jewish Civil Talent)	Monetary	327,783	1/60 5,463	1/4 218.5	1/4 54.6
Greek-Asiatic and Persian	Ditto	501,551	1/100 5,015	1/25 200.6	1/4 50.1
Attic from Olympic	Commercial	452,553	1/50 9,051	1/25 362	1/4 90.5
Euboic and Attic	Monetary	393,525	1/60 6,558	1/25 262.4	1/4 65.6
Egypto-Roman (from Great Alexandrian Talent)	Commercial	655,566 (100 lbs.) 1 lb = 5,244	1/12 437	—
Roman (from Greek-Asiatic)	Ditto and Monetary	501,551 1 lb. = 5,015	1/12 418	—
Carthaginian or Bosphoric (Mina, afterwards Yousdrouman pound of the Arabs)	Monetary	453,993	1/80 5,666	1/25 226.6	1/4 56.6
Arabian or Mahometan (based on Michtal, 1/10 Egypto-Roman pound)	Ditto	Rotl or 1 lb. = 7,283	369	1/10 = 72.8 Michtal
Ditto, Almamoun's system	Ditto	Canthar = 725,777 gr. " " 1/10 = 7,237		1/10 = 72.4 Michtal
	Ditto (Unit Cologne Marc of 8/9 lb. or 3,610 gr.)	" " 1/10 = 5,790		1/10 = 48.3 Dirhem
Old German (from Arab Rotl of 100 Michtals)	...	1 lb. = 7,220	—	—
Ditto, Medicinal	...	5,400	oz. 1/12 = 450	—
Frankish (from Arab pound sent by Almamoun to Charlemagne)	Monetary (Livre Esterlin)	5,666	oz. 1/12 = 472	—
Old French (Poids de Marc)	Commercial (Unit Marc = 1/2 Livre or 3,777 Eng. gr.)	7,554	oz. 1/16 = 472	—
Anglo-Saxon (old pound sterling of silver)	Monetary	5,400	S. 1/20 = 270	1/12 = 22.5 d.
Ancient English (Merchant's)	Commercial	6,750	oz. 1/15 = 450	—
Troy	Monetary	5,760	oz. 1/12 = 480	1/20 = 24 dwt.
Avoirdupois	Commercial	7,000	oz. 1/16 = 437.5	—
Metric	Monetary & Commercial Kilogramme=	15,432	—	1/1000 = 15.4 grm.

WEIGHING AND MEASURING.

STANDARD UNIT MEASURES OF CAPACITY.

		Equiv. in Imp. Gall
Egyptian Lesser Artaba	Cube of Egyptian foot	6·464
,, Royal Artaba	,, ⅔ Royal Cubit	9·440
,, Grand Artaba of Sephad	,, Natural or Olympic Cubit	21·818
Hebrew Ephah or Bath	,, ⅔ ditto	6·468
,, Hin, ⅙ of Bath (for liquids)	...	1·078
,, Gomer, 1/10 of Hin (for grain)	...	0·646
Syro-Phœnician and Persian Cafiz or Metretes	Cube of ¼ Chaldæan Cubit...	7·186
Syro-Phœnician Artaba (for grain)	Double the Cafiz	14·372
Arab Woebe	Half the Cafiz	3·593
Greek Metretes (for liquids)	Cube of Olympic Foot	6·464
,, Medimna (for grain)	1¼ Metretes	8·615
Roman Amphora (for liquids)	80 lbs. weight of wine	5·725
,, Quadrantal (for grain)		
,, Modius (for grain)	⅓ of Quadrantal	1·908
Old English Winchester Bushel	2,150·42 Cubic Inches	7·754
,, Queen Anne's Wine Gallon	231 ,, ,, ,,	0·834
Imperial Standard Bushel	{ 2,218·191 ,, ,, or 80 lbs. weight of water }	8·000
,, ,, Gallon	{ 277·274 Cubic Inches, or 10 lbs. weight of water }	1·000
Metric Litre (for liquids)	Cubic Decimetre	0·220
,, Hectolitre (for grain)	100 Litres	22·018

III. *English Standard Units of Weights and Measures.*

35. The English standard units of measures of length and capacity and of weight, the yard, bushel, and pound, have come down to us from the Saxons, though some modifications of the two last-mentioned units have since been made. No change was made by the Normans in the Saxon system of weights and measures established in England; and by a statute of William the Conqueror it was ordained that "the measures and weights shall be true and stamped in all parts of the country, as had before been ordained by law." The only change that appears to have been made at the Conquest was in the custody of the

standards, which was transferred from the City of Winchester to the Exchequer at Westminster, where they were placed under the custody of the king's chamberlains, whose office was part of the ancient Exchequer. The Exchequer itself was a Norman institution, and even up to the present time there may be seen at Rouen the ancient building of the Norman Exchequer, with the inscription upon it, " Le Père de l'Exchiquier de Londres." The English standards of weight and measure were deposited by the king's orders in a consecrated building, just as the standards of ancient countries were placed in their temples. Together with the royal treasures, they were placed in the crypt chapel of Edward the Confessor in the cloisters of Westminster Abbey, since known as the Pyx Chapel or chamber, from the standard trial plates for gold and silver coin used at trials of the pyx, also entrusted to the king's chamberlains, being kept in the same place of deposit. This portion of the old Abbey of Westminster then became vested in the sovereign, and has ever since been held by the officer who has had charge of the standards. Some of the old standards not required for actual use, and including the standard trial plates, continued to be deposited in the Pyx Chamber up to a recent period ; whilst those in actual use were kept at the Exchequer. In 1866, when the Exchequer ceased to be a separate office of the government, and was amalgamated with the audit office, the Standards Department of the Board of Trade was created, and all the standards passed to the custody of the warden of the standards, when they were deposited in the

E

strong fire-proof room of the standards' office, and kept in iron fire-proof chests.

36. The earliest recorded standard of length in this country was the yard or *gird* of the Saxon kings kept at Winchester. King Edgar is recorded to have decreed, with the consent of his wites, or council, that "the measure of Winchester shall be the standard." The yard and the ell were originally identical measures in England. From the period of the Conquest down to the time of Richard II., the statutes and official documents were either in Latin or in Norman-French, and the yard and ell (*virga* or *verge*, *ulna* or *aulne*) are employed indiscriminately to indicate the same unit of length. In the clause of Magna Charta relating to weights and measures, the term *ulna* is used as the unit of cloth-measure, whilst in Doomsday Book land measured by the yard is called *terra virgata*. The identity of the yard and ell is also clearly established by the old statute of uncertain date, prior to the reign of Edward II., entitled "Compositio ulnarum et perticarum," in which we find the well-known rule laid down that three barleycorns make an inch; twelve inches a foot; three feet an ell (*ulna*); five-and-a-half ulne a perch; forty poles in length and four in breadth an acre.

Another old statute of 1439 worthy of notice, 28 Henry VI. c. 16, is thus translated in the statute-book: "There shall be but one measure of cloth throughout the realm, by the yard and the inch (*la alne et le pous*), and not by the yard and handful (*l alne et la pleyne mayn*), according to the London measure." Later Acts down to 5 and 6 Edward VI.

ENGLISH STANDARDS.

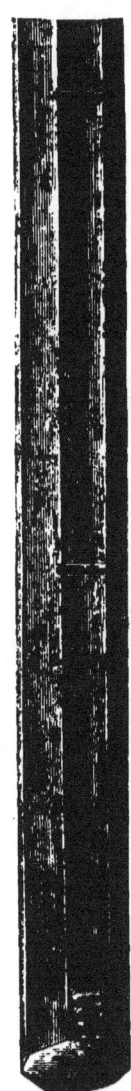

Fig. 8.—Exchequer Standard Yard of Henry VII. (Left-hand End), with Sixtenths of Yard.

Fig. 9.—Exchequer Standard Yard of Henry VII. (Right-hand End), with Sub-divisions of Inches.

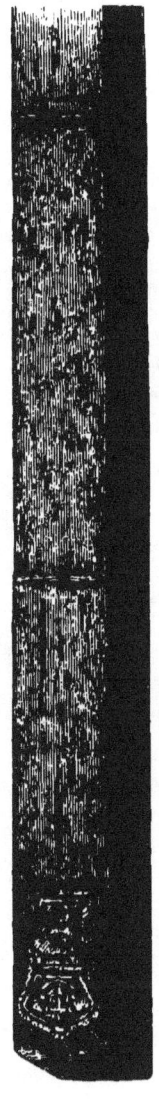

Fig. 10.—Exchequer Standard Yard Rod of Queen Elizabeth (Left-hand End), with Sixteenths of Yard.

c. 6, continue the measure of cloth "to every yard

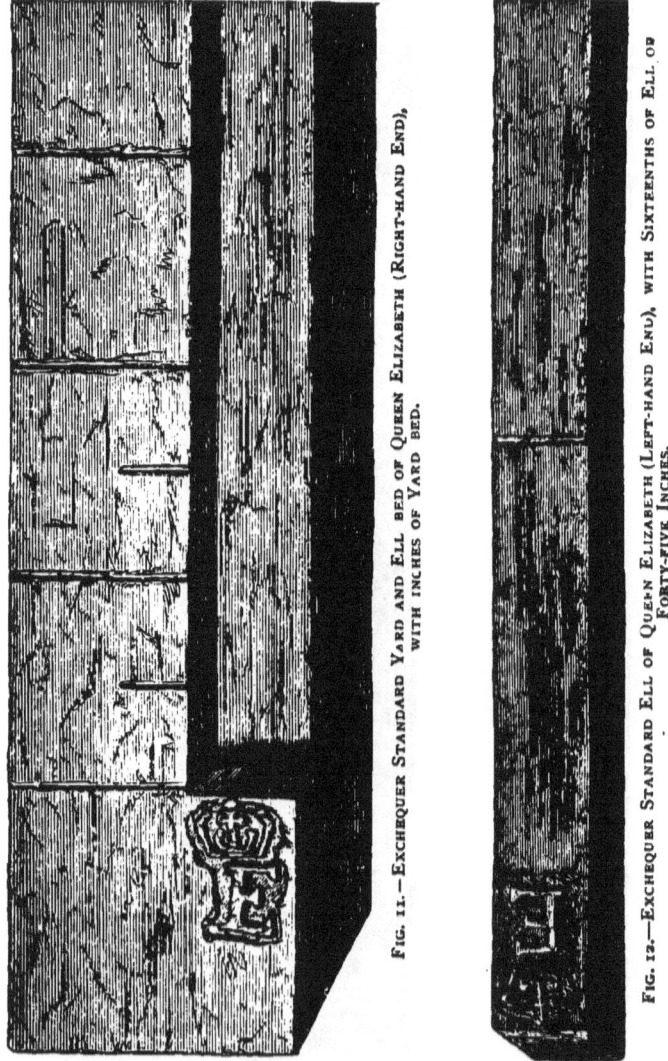

Fig. 11.—Exchequer Standard Yard and Ell bed of Queen Elizabeth (Right-hand End), with inches of Yard bed.

Fig. 12.—Exchequer Standard Ell of Queen Elizabeth (Left-hand End), with Sixteenths of Ell or Forty-five Inches.

one inch containing the breadth of a man's thumb,"

or "to every yard one inch of the standard." No such provision is contained in any later cloth Act, and it should be observed that the oldest English standard of length now existing consists of the exchequer yard of Henry VII. alone.

On the other hand, the exchequer standard yard constructed in Queen Elizabeth's reign has also the cloth ell of forty-five inches marked on the same bronze bar, both being bed measures, into which the standard yard and ell rods respectively fit. It is remarkable that although the standard ell itself exists, there is no mention of an ell of forty-five inches in any statute, nor in the existing records of the standards of Henry VII. and Queen Elizabeth. There is no sensible difference in the length of the Exchequer standard yards of these two Tudor sovereigns.

The existing imperial yard is so nearly identical in length with these old standard yards of Henry VII. and Queen Elizabeth, that it exceeds them by little more than a hundredth part of an inch, a difference frequently found in foot-rules now commonly used. There can also be but little doubt that our imperial yard is substantially the same length as the old Saxon yard. We have no further direct trace to its origin. But the English yard is so nearly the same length as double the natural cubit of the Egyptians and Hebrews, and the English foot is so nearly identical with the ancient foot, $\frac{2}{3}$ of this cubit, that the origin of these two English units of length may not improbably be traced to these two earliest standard units. We know that the double cubit was used in ancient times as a measure of length. An old Egyptian double royal cubit, found in the ruins at Karnac, may be seen in the British Museum. We know also that a measure very nearly equal to two natural cubits was used by the Romans under the name of *ulna*, or ell. The *ulna* is mentioned by Pliny, when describing the measurement of the girth of a tree, as half the length of the extended arms of a man. It may thus be fairly assumed that the measure of the double natural cubit, or three feet, under the name of ell or yard, came into use in old times, as a very convenient measuring unit, and found its way into England as the standard unit of length.

37. The earliest standard of monetary weight in England was the old pound of the Saxon moneyers in use before the Norman Conquest. The only legal standard of this pound of which any account has come down to us was the mint pound at the Tower of London, known as the Tower pound. It was of the same weight as the old apothecaries', or medicinal pound of Germany, and was equal to 5,400 of our later troy or imperial grains, and this weight of silver coins was the earliest form of our pound sterling. Both the Tower pound and the German medicinal pound are supposed to have been derived from the Ptolemaic mina, the sixtieth part of the lesser Alexandrian talent of silver. In 1842, an ancient weight of brass was found in the Pyx Chamber that weighed 5,409 troy grains, evidently an old monetary pound, somewhat increased in weight from oxidation. The pound sterling of silver was divided into twenty shillings, each of twelve pence, or pennyweights, as they were afterwards called. This scale of money

weight was the same as that of the old livre esterlin of Charlemagne, the livre being equal to 5,666 troy grains. Paucton quotes from an author contemporary with Charlemagne,—"Juxta Gallos, vicesima pars unciæ denarius est, et duodecim denarii solidum reddunt, ideoque juxta numerum denariorum tres unciæ quinque solidos complent." The old Exchequer records show that down to the end of the reign of Edward III., the weight of all gold and silver articles in the king's treasury was expressed in pounds, shillings, and pence, *pois de la Towre*, or *poise d'orefevre*. The pennyweight contained thirty-two grains, and thus the Tower pound contained 7,680 monetary grains, or grains of wheat. An Act of the 51 Henry III., stat. I. A.D. 1266, recites, as translated in our statute-book, "that by the consent of the whole realm of England, the measure of our Lord the King was made, viz., an English penny, called a sterling, round and without any clipping, shall weigh thirty-two wheatcorns in the midst of the ear; and twenty pence do make an ounce, and twelve ounces a pound; and eight pounds do make a gallon of wine, and eight gallons of wine do make a bushel, which is the eighth part of a quarter." This pennyweight was equal to 22½ troy grains, which is found to be the average weight of existing coined silver pennies of the Saxon and Norman kings. The mark was two-thirds of the Tower pound, and was also used for denoting both the weight and value of silver under the Norman kings. It was equal to 3,600 troy grains, and did not sensibly differ from the ancient unit of money weight in Germany, the Cologne mark.

38. The Tower pound was abolished as a legal mint weight in 1527, by an Ordinance of 18 Henry VIII., enacting that "the Pounde Towre shall be no more used and occupied, but al maner of golde and sylver shall be wayed by the Pounde Troye, which maketh xii oz. Troye, which excedith the Pounde Towre in weight iii quarters of the oz." From this time up to the present, the weight of our coinage and of the precious metals has been regulated by the troy system. Troy weight had, however, been introduced into England long before the reign of Henry VIII., and was certainly in general use there in the early part of the reign of Henry V., being mentioned in the Act 2 Henry V. c. 4. In the Exchequer records, the latest use of the old Tower money weight is in the 7th Richard II., after which date the weights are generally stated in pounds and ounces, but the term "pois de troie" appears to have been first used in an inventory of the king's gold and silver plate in the first year of the reign of Henry IV. It is probable that troy weight was brought into England from France during the wars and the English occupation under the Black Prince. Its name has been considered to have been derived from the French town of Troyes, where a celebrated commercial fair was held. There was a known *livre de Troyes* and *marc de Troyes*, and the recorded weight of the marc de Troyes was very nearly two-thirds of our pound troy.

39. The grain does not appear to have formed part of the English system of weights until some time after the Conquest, and to have been introduced from France, as the French *denarius* or *denier* contained

twenty-four grains. The old French grain esterlin was $\frac{1}{3780}$th part of the old French pound of Charlemagne, and was equal to 0·985 troy grain. Although the penny, or pennyweight, of silver of the old English money weight is declared in the ancient statutes of Henry III., Edward I., and Henry VII. to have been of the weight of thirty-two grains of dry wheat taken from the middle of the ear, the wheat grain being thus equal to 0·703 troy grain, there is no evidence of any such metallic grain-weights having been in use; and until the introduction of troy weight, the smallest weight used would appear to have been the farthing of silver, equal to 5·625 troy grains. But upon the introduction of troy weight into England, and of a gold coinage in the reign of Edward III., metallic grain-weights appear to have been established, and to have been practically used. Together with the ancient moneyer's pound which was found in the Pyx Chamber, there was also discovered a small neatly-turned box with an inscription upon it in the handwriting of the period, certainly not later than the reign of Edward III.,—" Grana pro auro." It contained a small disc of copper, marked with two dots, and was evidently a two-grain weight. It is but little oxidised, and probably differs little from its original weight. Its weight at the present time is 1·869 grain of our present standard, one grain being equal to 0·934 troy grain.

40. The use of both the Tower pound and the troy pound was confined to the precious metals and drugs. The commercial or merchant's pound (*libra mercatoria*) was also in general use at a very early period, and

is mentioned in a statute of 54 Henry III. as a pound of twenty-five shillings (una libra, pondus viginti-quinque solidorum legalium sterlinorum), to be used for all other commodities than the precious metals and medicines. It consequently contained fifteen tower ounces, or 6,750 troy grains. This commercial pound was thus one-fourth greater than the money pound. There was also a commercial pound, which was much used in France and Germany, and even in England, differing little from 7,200 troy grains, being thus one-fourth greater than the troy pound, and equal to sixteen tower ounces.

The merchant's pound, whether of 6,750 or 7,200 grains, appears to have been generally superseded by the avoirdupois pound of sixteen avoirdupois ounces, and equal to 7,000 troy grains, as early as the year 1303, it being recited in a weights and measures statute of that year, 31 Edward III. The word "avoirdepois," as applied to commodities, occurs in earlier statutes of the 9th and 27th of Edward III. The pound avoirdupois was evidently taken from the old French commercial pound of sixteen ounces, which was used in many parts of France. Our existing avoirdupois pound can be clearly proved to be of similar weight to the standard avoirdupois pound of Edward III., and there is good ground for believing that no substantial difference has occurred in its weight, or in that of the troy pound, since either of them was first established as a standard in this country. There is no evidence of any direct relation between the two units, the pound troy and the pound avoirdupois, when first introduced and legalised, pro-

ENGLISH STANDARDS. 59

bably at the same period, in this country. But we know from the Elizabethan standards that the pound avoirdupois was equal to about 7,000 grains, of which the troy pound contained 5,760 grains.

41. As regards the actual standards of avoirdupois weight, official records show that the series of Exchequer standard avoirdupois weights constructed in the reign of Queen Elizabeth, by which all the

FIG. 13.—STANDARD 14-LB. AVOIRDUPOIS WEIGHT OF HENRY VII. (⅓ size), EXETER.

commercial weights of this country were regulated up to the reign of George IV., and then remaining in the Exchequer, were derived from a 56-lb. avoirdupois standard of Edward III. But no trace could be found of this old standard in 1758, when inquiry was made for it by the Weights and Measures Committee of the House of Commons. New standard weights had, however, been constructed in the twelfth year of Henry VII., and in the Act, 12 Henry VII., which legalised them, it is expressly mentioned that the king

60 WEIGHING AND MEASURING.

"did make weights and measures according to old standards thereof remaining within his treasury." The original Exchequer standard weights of Henry VII. were destroyed when new standards were legalised

FIG. 14.—EXCHEQUER STANDARD 112-LB. OF QUEEN ELIZABETH (⅛ size)

by Queen Elizabeth, but some of the copies distributed to the several counties and principal cities have remained up to the present time. Figure 13 represents the brass 14-lb. avoirdupois standard weight of the set supplied to the city of Exeter, and formerly kept in the Guildhall there. It is now deposited in the Albert Museum in that city.

A portion of a series of avoirdupois standard weights, constructed in the seventeenth year of Queen Elizabeth's reign, was recently found deposited in the old Pyx Chamber. This set had, how-

FIG. 15.—EXCHEQUER STANDARD TROY POUND OF QUEEN ELIZABETH, FORMED OF 8-OZ. AND 4-OZ. WEIGHTS (actual size).

ever, been condemned as inaccurate at a later period of her reign, and it is evident that they had never been properly adjusted. These weights are of gun-metal, with iron rings, and are distinguishable from the later and more accurate set of Queen Elizabeth's avoirdupois standards by being marked with a rose. They are of elegant form, as may be seen from fig. 14, representing the 112-lb. weight.

The more accurate and complete set of Exchequer standard weights, constructed in 1588 under Queen

Elizabeth's orders, and made of bell-metal, are still in good condition. They were used to regulate all the weights in the kingdom until 1824, when the new imperial standard weights, constructed under Captain Kater's directions, were substituted for them under 5 Geo. IV. c. 74. The set of troy standards

FIG. 16.—EXCHEQUER STANDARD AVOIRDUPOIS POUND OF QUEEN ELIZABETH, BELL-SHAPED (actual size).

from 1 oz. to 256 ozs. are in the form of cup weights fitting into each other. There is no single weight of a pound troy, but the Exchequer standard of the troy pound was formed of the two weights of 8 oz. and 4 oz. as shown in fig. 15.

There are two sets of avoirdupois standards. The larger set, from 56 lbs. to 1 lb., is bell-shaped, as shown in fig. 16. This avoirdupois pound is believed to have been originally equal to about 7,002 troy grains. In 1758 the Weights and Measures Committee of the House of Commons reported it to weigh 7,000·5 troy grains. In 1873 it weighed 6,999 grains of the

FIG. 17—EXCHEQUER STANDARD AVOIRDUPOIS POUND OF QUEEN ELIZABETH, FLAT DISC SHAPE (actual size).

imperial standard. It thus appears to have lost only three grains in weight, though continually used as an Exchequer standard from 1588 to 1825.

The smaller set of Elizabethan avoirdupois standards, from 8 lbs. to 1 dram in a continued binary series, consists of flat circular weights, as shown in fig. 17. This avoirdupois pound was found in 1758 to weigh 6,997·5 troy grains. In 1873 it was found equal to 6,996·4 grains of the imperial standard.

The bell shape, with its convenient handle, has

FIG. 18.—FORM OF MODERN LOCAL STANDARD AVOIRDUPOIS WEIGHTS (actual size).

generally been preferred for the larger avoirdupois standard weights, both for lifting and carrying them over the shoulder. The form shown above is that which has been generally adopted for local standard weights.

42. As to our early English standard measures of capacity, the first of which we have any cognizance are the Winchester corn-bushel, of a capacity of 2,150½ cubic inches, and the Winchester corn-gallon, of 274¼ cubic inches. The oldest standards of these measures that now remain in the Standards Office are those constructed by order of Henry VII. (Figs. 19 and 20).

FIG. 19.—EXCHEQUER STANDARD WINCHESTER BUSHEL OF HENRY VII. (⅛ size).

A standard ale gallon of 282 cubic inches was added by Queen Elizabeth in her new set of standard measures constructed in 1601, the standard quart of which is here shown (Fig. 21). In 1707 another addition was made to the Exchequer standards by Queen Anne, of a standard wine-gallon of 231 cubic inches (Fig. 22). But all these standard measures of capacity

F

were abolished in 1824, when the new imperial standard gallon, containing 10 lbs. weight of water

FIG. 20.—EXCHEQUER STANDARD CORN GALLON OF HENRY VII. (⅕ size).

and of the capacity of nearly 272¼ cubic inches, was made the standard of capacity for liquid measures, and the imperial standard bushel of eight gallons was made the standard for measuring grain and dry commodities.

43. All the before-mentioned old English standards were brought under the notice of the Weights and Measures Committee of the House of Commons, who sat in 1758 and the two following sessions. Many recommendations were made by this committee in relation to our standards, which were not, however,

Fig. 21.—Exchequer Standard Quart of Queen Elizabeth (½ size).

then carried into effect by law. Under their directions a new standard yard and standard troy pound, both of brass, were constructed, and were intended to be constituted the legal standards of length and weight in this country. Meanwhile they were left in the custody of the Clerk of the House of Commons. But it was not until 1824 that the new imperial system of weights and measures was legalised, in the Act 5 Geo. IV. c. 74. In this Act the standard yard of 1760, and the brass troy pound of 1758, both in the custody of the Clerk of the House of Commons, and denominated respectively the "Imperial Standard Yard" and the "Imperial Standard Troy Pound"—were declared to be the only "original

FIG. 22.—EXCHEQUER STANDARD WINE GALLON OF QUEEN ANNE (⅓ size).

and genuine standards," from which all other imperial weights and measures were to be derived. No provision was contained in the Act of 1824 for removing the two primary standards from the House of Commons, and the result was that they were destroyed by the burning of the Houses of Parliament on 16th October, 1834. The lost standard pound was similar to that represented in Fig. 23, showing the original standard troy pound of 1758, now in

FIG. 23.—STANDARD TROY POUND OF 1758 (actual size).

the Standards Office, from which the lost standard had been constructed.

IV. *The Restored Standards.— Imperial Standard Pound and Yard.*

44. The legal standard units of imperial weight and measure are now the standard pound avoirdupois and the standard yard constructed under the superintendence of the Standards Commission, appointed in 1843, for the restoration of the lost standards of weight and measure.

The members of this Standards Commission had previously given their services as a preliminary committeee, having been appointed in 1838 to consider the steps to be taken for restoring the standards; the Act of 1824 (5 Geo. IV. c. 74), under the authority of which the lost standards had been legalised, having directed that, in the event of their loss or destruction, new standards should be constructed in accordance with provisions contained in the Act, by reference to an invariable natural standard.

These provisions were as follows:—In regard to the Standard of Weight, it was recited in § 5 of the Act, that a cubic inch of distilled water, weighed in air against brass weights, at the temperature of 62° Fahr., the barometer being at thirty inches, had been determined by scientific men to be equal to 252·458 grains, of which the standard troy pound contained 5,760; and if this standard were lost or destroyed, a new standard troy pound was to be constructed bearing the same proportion to the weight of a cubic inch of water, as the standard pound bore to such cubic inch of water.

It will thus be seen that the new unit of weight was to be dependent upon the new unit of length, as it was to be based upon the capacity of the cubic inch, or the cube of the thirty-sixth part of the standard yard.

With respect to the standard unit of length, § 3 of the Act recited that the imperial standard yard, when compared with a pendulum vibrating seconds of mean time in the latitude of London, in a vacuum at the level of the sea, had also been determined to be in the proportion of 36 inches to 39·1393 inches, and it was provided that if lost or destroyed, a new standard yard should be constructed bearing the same proportion to such pendulum as the imperial standard yard then bore to it.

After long deliberation, the committee made a very full report, dated December 21, 1841, and declared their opinion that the several elements of reduction of the pendulum experiments referred to in the Act of 1824, were doubtful or erroneous. It was evident, therefore, that the course prescribed by the Act would not necessarily reproduce the standard yard. It appeared also that the determination of the weight of a cubic inch of water was still doubtful, differences being found between the best English, French, Austrian, Swedish, and Russian determinations, amounting to about $\frac{1}{1200}$ of the whole weight, whereas the results of the mere operation of weighing might be determined within $\frac{1}{1000000}$ of the whole weight. The committee were fully persuaded that with reasonable precautions it would always be possible to provide for the accurate restoration of standards by means of material copies which had been compared with them. They had ascertained that several measures existed which had been most carefully compared with the former standard yard, and several weights, which had been most accurately compared with the lost standard pound, and they expressed their opinion that by the use of these the values of the original standards could be restored without sensible error.

They recommended that no change should be made

in the values of the primary units of the weights and measures of the kingdom, or in the meaning of the names by which they were commonly denoted; that the construction of the standards be entrusted to a committee of scientific men, under certain instructions contained in the report, and by comparison with the most carefully selected specimens; that the parliamentary standard of length be one yard, there appearing no sufficient reason for departing from the length hitherto adopted for the standard; and that the avoirdupois pound be adopted instead of the troy pound as the parliamentary standard of weight, the avoirdupois pound being invariably known and generally used, and the troy pound being wholly unknown to the great mass of the British population, and comparatively useless. They also recommended that no new specific standard of capacity be established, the unit of capacity, the gallon, being continued to be defined by its containing 10 lbs. weight of distilled water, as specified in the Act of 1824.

Many other important recommendations were also made by the committee relating to the official secondary standards, and to the verification and legalising of local standards for the use of inspectors of weights and measures throughout the country and for the colonies; as well as to the regulation of the duties of these local officers, in order to secure due uniformity in commercial weights and measures, and their accordance with the scientifically constructed primary standards.

For more effectually carrying out the recommendations for the construction of the new standards, the Standards Commission was appointed in 1843, and continued their labours until 1854, when they presented their definitive report.

The Astronomer-Royal, Mr. Airy (now Sir G. B. Airy, K.C.B.), was chairman both of the preliminary committee and of the commission. The other nine members of the commission were all eminent men of science. The mode in which the commission performed their duties in the restoration of the standards, and the results of their labours, as shown in their definitive report, will next be described.

Construction of New Imperial Standard Pound.

45. The mode of constructing the new standard of weight, together with full details of all the scientific processes employed, has been described by Prof. W. H. Miller, the member of the commission to whom its construction was more immediately entrusted. (See *Phil. Trans.* 1856, Part III.)

46. For constructing this standard, the first point to be determined was the exact weight and density of the lost standard troy pound, from which the weight of the new standard avoirdupois pound was to be derived. Upon investigation this proved to be the most difficult problem to be solved by the commission. The old standard had been constructed in 1758, under the direction of the parliamentary committee of that year, together with three similar troy pounds. It is stated to have been composed of gunmetal, but unfortunately no record existed of its

volume or density, and it is not probable that it was ever weighed in water. An accurate drawing of the lost standard pound had been made in 1829 by Captain Nehus, who measured its dimensions with the greatest care. (See *Phil. Trans.* 1836, p. 361.) From a comparison of this drawing and the dimensions with the three other troy pounds, the density of the lost standard was assumed by Prof. Miller to be 8·151.

The three other troy pounds constructed in 1758 were found by the commission to differ slightly in their dimensions, as well as in volume and weight. They were all in good preservation and were carefully examined by Prof. Miller, but there was no satisfactory evidence of their having been accurately compared with the lost standard, so as to identify its weight, and thus to render them available for determining the proper weight of the new standard. The troy pound of 1758, now in the Standards Department, was computed by Prof. Miller to weigh in air 5759·85625 grains of the lost standard.

For ascertaining the exact weight of the lost standard pound, the following weights, which had been accurately compared with it, were also examined:—

>The brass troy Exchequer standard pound, constructed in 1824 under the superintendence of Captain Kater, and legalised as the secondary official standard;
>
>Three similar brass pounds, constructed for the cities of London, Edinburgh, and Dublin;
>
>A platinum troy pound and two brass troy pounds belonging to Prof. Schumacher;
>
>The platinum troy pound of the Royal Society.

47. It was found, however, from examining the results of several weighings of the brass troy pounds, that great discrepancies existed, attributable to the effect of oxidation or other causes. It was consequently resolved to rest entirely for evidence of the weight of the lost standard on the comparisons of the two platinum troy pounds of Prof. Schumacher and of the Royal Society, denoted as Sp and RS. These two platinum weights had been constructed in 1829, and were intended to be equal to the lost standard (denoted as U) when weighed in air. Each of them had been compared with U by Captain Nehus at Somerset House in 1829, with the following results:—

Mean of 300 observations, Sp = U — 0·00857 grain (mean
$t = 65°·62$ F. $b = 29·722$ in.)
Mean of 140 observations, RS = U — 0·00205 grain (mean
$t = 65°·73$ F. $b = 29·806$ in.)

The density of Sp had been determined, by weighing it in water, to be 21·1874; and it was found to displace 0·32544 gr. of air of the stated mean temperature and atmospheric pressure. The density of U was assumed to be 8·151, which is nearly the average density of brass and bronze weights, and U therefore displaced 0·84646 gr. Whence in a vacuum Sp = U − 0·52959 gr.

The density of RS also had not been determined by weighing in water, but it was assumed to be of the same density as Sp, and therefore to have displaced 0·32629 gr. of air, whilst U displaced 0·84865 gr. Whence in a vacuum RS = U − 0·52441 gr. The mean value of the lost standard troy pound thus determined through Sp and RS, was the basis upon which the new standard avoirdupois pound was to be constructed.

48. As a preliminary operation, a new platinum troy pound, denoted as T, was constructed very nearly equal to Sp and to RS, and taking the mean

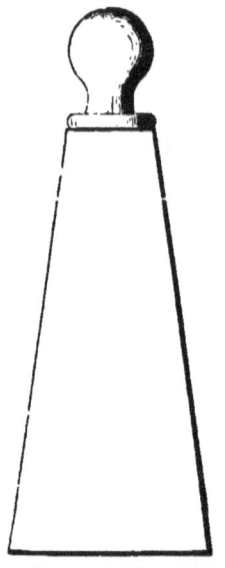

FIG. 24.—PLATINUM TROY POUND OF THE ROYAL SOCIETY [R S] (actual size).

FIG. 25.—PLATINUM TROY POUND OF THE STANDARDS DEPARTMENT [T] (actual size).

of 286 comparisons of T with Sp, and of 122 comparisons of T with RS, it was found that in a vacuum

$$T = Sp + 0\cdot 00105, \text{ whence } T = U - 0\cdot 52851$$
$$T = RS - 0\cdot 00429, \text{ whence } T = U - 0\cdot 52870.$$

From the mean of these two results, giving to the first twice the weight of the second, in consequence of Sp having been compared about twice as many times with U and with T as RS was compared, it was finally determined that in a vacuum

$$T = U - 0{\cdot}52857 \text{ gr., or} = 5759{\cdot}47143 \text{ grs.}$$

It was also found that in air, $t = 65° 66$ F. $b = 29{\cdot}75$ in., which was the mean of the comparisons of Sp and RS with U, and was adopted by Prof. Miller as the standard air, $\quad T = U - 0{\cdot}00745$ gr.

The form and size of the two platinum troy pounds RS and T are shown in Figs. 24 and 25; RS being a truncated pyramid surmounted with a knob, and T a cylinder with a groove.

It should here be observed that all the standard troy pounds were intended to be of their true weight in ordinary air, whilst the new standard imperial avoirdupois pound was to be made of its true weight when weighed in, or reduced to, a vacuum.

The next process was to construct the new avoirdupois standard pound, and to accurately adjust its weight of 7,000 grains in a vacuum, from the troy pound T. For this purpose four new platinum weights of 1,240 gr. each were constructed, all accurately verified in terms of T, and by employing other platinum weights, viz. one each, of 800 440, and 360 gr., four of 80, and two of 40 gr., the true weight in a vacuum of each of the 1,240 gr. weights, as separately determined by numerous comparisons with T and with each other, was as follows:—

```
              Grains.
      A = 1239·88622
      B = 1239·88605
      C = 1239·88597
      D = 1239·88580
      Mean = 1239·88601
T + Mean = 6699·35744
```

It thus required only a weight of 0·64266 gr. to make up the full weight of 7,000 gr. The approximate weight of 0·645 gr. was obtained from T in the following manner. By comparisons with the two 40 gr. platinum weights, two platinum weights of nominally 20 gr. each were found to weigh 19·998 gr. each, from which were derived $W = 12·901$ gr., $V = 6·451$ gr. From V was derived $Q = 0·64509$ gr., the mean of ten weights of platinum wire, each equal to 0·645 gr. very nearly. It will be shown hereafter, in describing the mode of weighing with a scientific balance, that small differences between two standard pound weights of less than 0·1 gr. are ascertained by the index scale of the balance. Means were thus afforded of determining the exact weight of 7,000 gr., which was to be the weight in a vacuum of the new standard.

49. The new imperial standard pound was constructed of platinum, and denoted as PS or Parliamentary Standard. Its form and size have been already shown in Fig. 1.

The true weight of PS was actually determined from the mean result of 340 comparisons with the following platinum weights, viz. :—

$$PS = T + Q + \tfrac{1}{4}(A + B + C + D) - 0·00177 \text{ gr. in air } t = 19·28° C. \; b = 759·12 \text{ mm.}$$

The density of PS was determined by weighing in water to be 21·1572, and that of T and the smaller platinum weights to be 21·1661. PS consequently displaced 0·39744 gr. of air, and $T + Q + \tfrac{1}{4}(A + B + C + D)$ displaced 0·39727 gr. Hence

$$PS = 7000·00093 \text{ grains, of which U contained 5760.}$$

RESTORED STANDARDS.

Having arrived at this very close approximation to the desired weight of the new standard, it was resolved by the commission that PS should be constituted the new imperial standard pound, and be deemed to contain 7,000 grains of the new standard.

50. Four similar platinum pounds were constructed at the same time, and their weight in terms of the new standard PS accurately determined. These four pounds were intended as Auxiliary Standards of Reference, with the view that either of them might replace PS in case of its destruction or damage. They were termed Parliamentary Copies (PC), and were deposited as follows :—

PC, No. 1, at the Royal Mint.
PC, No. 2, with the Royal Society.
PC, No. 3, in the Royal Observatory at Greenwich.
PC, No. 4, immured in the New Palace at Westminster.

Thirty-six other standard pounds of bronze gilt were also constructed, and their standard weight, both in a vacuum and in the standard air adopted by Prof. Miller, was accurately determined, as well as the densities of all the new standard pounds. These gilt bronze pounds were distributed amongst different countries and public institutions of this country.

All the numerous weighings both in air and in water of the new standard pounds, for determining their weights and densities, were made by Prof. Miller himself, and full details of all these operations are given by him in his *Account of the Construction of the new National Standard of Weight.*

51. The new imperial standard pound is of the true weight of an avoirdupois pound when in a vacuum.

The principal advantage of the metal of which it is composed (platinum) consists in its not being affected by oxidation, which would unavoidably alter its absolute weight. But platinum has this disadvantage,* if used as the material of a standard for regulating ordinary weights of precision made of brass, viz. that when weighed in air against a brass or bronze standard weight of so much greater volume, although of equal weight in a vacuum, its apparent weight is always more than half a grain greater than that of the brass or bronze standard. To obviate this disadvantage, the weight in air of all the bronze standard pounds verified by Prof. Miller were computed by him, *not in terms of the platinum standard pound, but of an ideal brass commercial standard pound, denoted by him as W.* He assumed W to be of the same density as the lost standard, and of the average density of brass or bronze. In air, $t = 65°·66$, $b = 29·75$ in., PS with a density of $21·1572$ displaced $0·39644$ gr. of air, and W was assumed to displace $1·03051$ gr. And as the official standard weights, by reference to which all commercial weights are verified, are made of brass or bronze, it was intended that they also should be regulated by their weight in air when referred to the brass commercial standard W. This has in fact been done. The only change since made has been under the sanction of the late Standards Commission, by which the density of the standard air recited in the Act of 1824 for determining the weight in air of a cubic inch of water, viz. $t = 62°$ F., $b = 30$ in., has been substituted for that adopted by Prof. Miller from its being the air in which the weight of the lost standard pound had been most accurately determined. The object of this change was to adopt one uniform standard temperature and barometric pressure for all standard purposes. In the new standard air (log. $\Delta = 7\cdot0852825 - 10$), P S displaces $0\cdot40282$ gr., and W, with a density of $8\cdot1430$, displaces $1\cdot04706$ gr. of air.

Construction of New Imperial Standard Yard.

52. The immediate superintendence of the construction of the new standard yard was entrusted, in the first instance, to Mr. Baily, who conducted all the preliminary investigations and experiments. After his death in August, 1844, it was undertaken by another member of the Commission, Mr. Sheepshanks, by whom and under whose direction by far the largest proportion of the actual operations was carried out, and all the comparing operations of the several standards of length made, up to the period of his death in August, 1855. By this time the work was so far completed that not a single additional comparison of line measures was required. The detailed account of the construction of the new standard yard, and its verified copies, was then undertaken by the Astronomer-Royal, with the aid of the documents left by Mr. Baily and Mr. Sheepshanks; and the winding-up of the work of the Commission, and the distribution of the scientifically verified copies of the standards, also devolved upon the Astronomer-Royal, as the chairman. The magnitude of the operations may be estimated from the fact of the number

of micrometer readings for all the comparisons exceeding two hundred thousand ; amongst the operations, also, it was found necessary to construct an entirely new system of thermometers. It should not be forgotten that the scientific gentlemen who bestowed so much of their valuable time, attention, and labour, during several years, upon the experiments and observations for the important object of the restoration of the national standard of length, declined to accept any pecuniary remuneration.

The length of the new standard yard was determined in a similar manner to the determination of the weight of the new standard pound, by taking the mean length of the most authoritative standards which constituted the best primary evidence of the lost standard yard.

53. This standard had been constructed by Bird, in 1760, under the directions of the Committee of the House of Commons on Weights and Measures, first appointed in 1758. Its length was taken from a similar yard, which had been constructed by Bird in 1758. Each of these two standard yards consisted of a solid brass bar 1·05 inch square in section, and 39·73 inches long. Near each end of the upper surface gold pins or studs, 0·1 inch in diameter, were inserted, and points or dots were marked upon the gold to determine the length of the yard. The comparing apparatus in use at that period consisted of a beam compass with two fine measuring points, which could be adjusted to the dots on the standard measures under comparison. But the result of numerous comparisons of this kind made from time to time, previously

to the destruction of the standard in 1834, had been to leave the edges of the holes indented and irregularly worn away, so that the original centre was very difficult to ascertain. Mr. Baily, who had made some comparisons with this standard yard in the early part of the year 1834, describes the holes as appearing, under a microscope, like the miniature crater of a volcano.

The length of the standard yard of 1758 had been based upon that of the then existing Exchequer standard yard, which had been constructed in the reign of Queen Elizabeth in 1588, and upon the length of the Royal Society's standard yard, constructed as a scientific standard measure in 1742. It had been determined, upon comparison, to agree as nearly as possible with these two authoritative measures of a yard.

The two standard bars of 1758 and 1760 were both found amongst the ruins of the Houses of Parliament, but they were too much injured to indicate the measure of a yard which had been marked upon them.

Meanwhile, since the construction of Bird's standard yard in 1760, other scientific standards of length had been constructed. In 1785, the first geodesical operations were begun, upon which the Ordnance Survey of the United Kingdom has since been founded, by General Roy's measurement of the base on Hounslow Heath. The standard used in the first instance for that purpose was that known as General Roy's scale, 42' inches in length, and constructed by Mr. Bird. This scale was based, not on the legal Exchequer standard, but upon the Royal Society's standard scale of 42 inches, with the first 36 inches of which the same length only of General Roy's scale was compared, this constituting the *Ordnance yard.* Two standard yards of superior construction, belonging to the Ordnance Department, were placed at the disposal of the Standards Commission. These were bars of iron, and line standards, the lines being marked on gold pins at mid-depth of the bar, notches being cut in it for that purpose. They had been compared with the imperial standard in 1834, and a statement of their comparison was published in 1847 in the account of the measurement of the base at Lough Foyle.

Towards the close of the century some important scientific operations for the improvement of the standards were undertaken by Sir George Shuckburgh. In 1796 a new standard measure subdivided in fine lines, and since known as "Shuckburgh's scale," was constructed under his direction by Mr. Troughton, together with a new comparing apparatus carrying micrometer microscopes. This is stated to have been the first occasion on which such a mode of optical comparison was employed, being substituted for the beam compasses previously used. The Shuckburgh scale, which is now in the possession of the Royal Society, consists of a brass bar $67\frac{3}{4}$ inches long, 1·4 inch wide, and 0·42 inch thick. It is a scale of five feet, divided by lines into feet, inches, and tenths of inches, each inch being numbered. It was adopted by the Standards Commission of 1819 as the scientific standard of length, as distinguished from the legal standard of the Exchequer. The length of the yard was laid down on the Shuckburgh scale from Bird's standard, and it had also been accurately compared with each of the other standard yards previously mentioned, and their lengths had been transferred by beam compasses to the Shuckburgh bar.

In pursuance of the recommendation of the Royal Commission of Weights and Measures appointed in 1819, and of the Act of 1824, passed to carry their recommendations into effect, a new Exchequer standard yard for regulating commercial measures of length was constructed under Captain Kater's superintendence. It was not, however, laid down from the legal standard yard, which, together with the legal standard pound, remained in the custody of the Clerk of the House of Commons, but from the length 1—36 inches of the Shuckburgh scale, which was considered by Captain Kater to be identical with the imperial standard.

This Exchequer standard yard consists of a slender brass rod with two wooden handles, as an auxiliary end-measure, and of a bed measure, being a bar of brass one inch square, with rectangular steel terminations of the same width projecting above the surface of the bar. The distance between the interior faces of the steel terminations is intended to be equal to the length of the imperial yard. This yard bed and rod were used together, from 1825 to 1870, for verifying all the local standard yards of similar though ruder construction. A standard yard with the legal subdivisions marked upon it, and of improved construction, having a convenient comparing apparatus attached to it, has since been substituted, and is now used in the Standards Department.

Four other standard yards of more scientific character were also made under Captain Kater's directions, and are now in the Standards Office. These bars of brass are of the same width and thickness as the Shuckburgh scale, and have the length of the yard defined by fine points upon gold studs in the middle axis of the bar, the thickness of the bar being reduced at its extremities one-half with this object. All these standard yards were constructed by Dollond. By an ingenious contrivance the point at one end of the bar, not being placed exactly in the centre of the circular gold stud, was made susceptible of adjustment by turning the stud round; and after final adjustment of each yard, and repeated comparisons with the Shuckburgh scale, no perceptible error could be detected in any of them. A similar standard measure made for the Royal Society in 1831 was considered by the Commission to be the most favourable type of Kater's yard.

54. Having thus described the principal standard yard-measures then existing, we may return to the operations of the Standards Commission. For determining the true length of the new standard yard, a provisional standard yard was employed by Mr. Sheepshanks. This was a new brass bar, called by him "Brass 2," and was accurately compared by him with the standards deemed to be the most authoritative, and which had been *directly* compared with the lost standard, viz., Shuckburgh's scale, Kater's yard made for the Royal Society, and the two Ordnance yards. The results in terms of the lost imperial standard were as follows :—

RESTORED STANDARDS.

in.
Brass Bar 2 = 36·00084 by comparison with Shuckburgh scale 0–36 in.
,, = 36·000280   ,,   ,,   10–46 in.
,, = 36·000229   ,,   Kater's Yard of 1831.
,, = 36·000303   ,,   Ordnance Yard, No. 1.
,, = 36 000275   ,,   ,,   No. 2
──────────────────
,, = 36·000234 by mean of all.

Mr. Sheepshanks preferred 36·00025, as being sufficiently near the truth, and in constructing the new standard, he assumed as the basis of his proceedings—

Brass 2 = 36·00025 in. of lost imperial standard, at 62° Fahr.—and this conclusion met with the assent of the Commission.

55. In the construction of the new standard of length, the following decisions were made by the Commission :—

1. The length of one yard to be the standard unit of length.

2. After considering whether the measure of length should be defined by the whole length of the bar, that is to say, an *end-standard*, or by the distance between either two points or two lines marked upon the bar, a *line-standard* was adopted in preference.

3. For the material of the bar, gun-metal, or bronze, composed of

Copper ... ... 16 parts
Tin ... ... 2½ ,,
Zinc ... ... 1 ,,

was adopted after a series of experiments by Mr Baily, and was recommended by him as containing the properties most essential for the construction of a standard intended to last through many ages, viz., almost perfect immunity from rust, with proved elasticity and rigidity. The test-bar of this alloy.

when loaded at the centre with 5½ cwt. broke without bending.

4. The form of the standard to be a solid bar 38 in. long, and 1 in. square in section. The measure of a yard to be defined by the distance between two fine lines perpendicular to the axis of the bar, marked upon gold studs at the bottom of cylindrical holes drilled from the upper surface to the mid-depth of the bar.

The gun-metal, or bronze, thus adopted for the new standard, has since been known as "Baily's metal," and this designation is engraved upon the imperial standard yard.

In order to select the most perfect specimen for the new standard of length, forty line-standard yards were constructed of Baily's metal, and one of these was finally selected as the imperial standard, not only from its representing, with the greatest precision, the assumed length of the lost standard yard, but also from the clearness of its defining lines, and from its general good workmanship. The details of the numerous comparisons of all these new standard measures, including the determination of the rates of expansion of the several bars, may be found in the Astronomer Royal's account, Phil. Trans. 1857, Part 3. The new standard was declared to have the true length of an imperial yard when at the temperature of 62° Fahr., and the exact temperature at which each of the other standard measures had this true length of a yard was ascertained and engraved upon each of them. Four of the remaining yards, nearest in length to the new standard, were selected as

Parliamentary copies, having the true length of a yard at the respective temperatures of 61°·91, 62°·10, 61°·98, and 62°·16 Fahr. These were deposited in the same places as the Parliamentary copies of the standard pound already mentioned; and the rest were in like manner distributed amongst different countries and public institutions in this country.

Several other similar line-standard yards were also constructed for experimental purposes, and were accurately verified by Mr. Sheepshanks, and disposed of in like manner, viz. :—

> 4 Bronze, with different proportions of alloy.
> 5 Brass.
> 3 Swedish iron.
> 4 Low Moor iron.
> 4 Cast iron.
> 4 Cast steel.

Together with the following end-standards—

> 3 Baily's metal.
> 1 Swedish iron.
> 1 Low Moor iron.
> 2 Cast steel.

The defining terminations of these end-bars consist of a plug of agate, slightly conical and shrunk into a similar conical hole at each end of the middle axis of the bar. The ends of the bars are ground and polished in a spherical form, the centre of the spherical surface being the middle of the bar.

All the numerous comparisons of the standard yards were made by Mr. Sheepshanks in one of the lower cellars at Somerset House, under the apartments of the Royal Astronomical Society, where the new micrometrical comparing apparatus constructed for the purpose by Messrs. Troughton and Simms was fixed.

56. The Commission for restoration of the Standards having terminated their labours, recommended in their final Report that the new imperial standards of the yard and pound be deposited at the Exchequer Office, there to be preserved under such regulations as to Parliament might appear fitting. In expressing their adherence to the recommendation of the Committee of 1841 that no reference should be made to natural elements for the values represented by the standards of weight and measure, they also recommended that so much of the Act 5 Geo. IV., c. 74, as provided for the restoration of the standards in the manner therein provided, be repealed, and that the standards should in no way be defined by reference to any natural basis, such as the length of a degree of the meridian on the earth's surface in an assigned latitude, or the length of a pendulum vibrating seconds in a specified place. They considered the ascertaining of the earth's dimensions, and the length of the seconds pendulum in terms of the standard of length, and the determination of the weight of a certain volume of water in terms of the standard of weight, as scientific problems of the highest importance, to the solution of which they trusted that Her Majesty's Government would always give their most liberal assistance, but they did not urge them on the Government as connected with the conservation of standards.

57. These recommendations were carried into effect

by the Act of 1855, 18 and 19 Vict., c. 72, for legalising and preserving the restored standards of length and weight, sec. 1 of which repealed the provisions of the Act of 1824 concerning the restoration of the standards by reference to the pendulum and to the weight of a cubic inch of water. Under the provisions of the Act of 1855, the imperial standards were deposited, in 1855, in the Office of the Exchequer.

58. Upon the creation of the Standards Department of the Board of Trade, under the Standards Act, 1866, 29 and 30 Vict., c. 82, the custody of the imperial standards was transferred to the Warden of the Standards. Provision is contained in this Act for the comparison once in every ten years of the three Parliamentary copies of the imperial standards deposited at the Royal Mint, in charge of the Royal Society, and in the Royal Observatory Greenwich, respectively, with the imperial standards of length and weight, and with each other. Under this Act new scientific duties were also imposed upon the Standards Department, the Warden of the Standards being charged with conducting all such comparisons, verifications, and other operations with reference to standards of length, weight, or capacity, in aid of scientific researches or otherwise, as may be required.

59. In connection with the question of the derivation of a standard unit of length from a natural constant to be found in the ascertained dimensions of the earth, it may be added that Sir John Herschel, who was a member of the Commission for restoring the

Standards, has pointed out the fact of the length of the polar axis having been determined, from the combined results of all the scientific measurements of arcs of the meridian, to be equal to 500,482,296 inches of our imperial standard yard, and that if one five-hundred-millionth part of the polar axis were adopted as a new standard unit, to be called the "geometrical inch," it would differ from the imperial inch less than one-thousandth part of an inch; a difference so small as not to be measured by any ordinary method, and only by the aid of the nicest scientific instruments. For all "ordinary practical purposes," the geometrical inch would be identical with the imperial inch; whilst for high scientific measurements for astronomical purposes, it would connect by an unbroken numerical chain the small units with which mortals are conversant in their constructions and operations with the great features of nature, and more especially with those greater units in the measurements of the universe with which astronomy brings us in relation. It would also produce a more exact ratio between our units of length and weight, the avoirdupois ounce being nearly a "geometrical ounce," or one-thousandth part of the weight of a geometrical cubic foot of distilled water. That is to say, whilst the existing legal weight of a cubic foot of distilled water is 997·136 ounces, the weight of a geometrical cubic foot of water would be 998·1 ounces. And as the imperial half-pint is the measure of ten ounces of distilled water, the ratios of these units of length, weight, and capacity would thus be brought within such practical limits of precision as would meet every possible requirement of commercial exigency.

V.—*Secondary Imperial Standards, including Multiples and parts of Standard Units.*

60. A complete series of secondary standards was constructed and accurately verified for the Exchequer under Kater's superintendence in 1824. They were based on the imperial standard yard and the imperial standard pound, as the primary units; and their value was not affected by the legal substitution of the restored imperial standard yard and pound for the lost standards upon which they had been based. They include the imperial standard gallon, the standard unit of measures of capacity, derived from the imperial standard pound, as containing 10 lbs. weight of distilled water, weighed against brass weights in air at the temperature of 62° Fahr., the barometer being at 30 inches. These secondary standards, together with others subsequently legalised, have served for regulating all the commercial weights and measures of Great Britain and her colonies and dependencies since 1824, when they were substituted for the old set of Exchequer standards which had been in force from the reign of Queen Elizabeth. All the Exchequer secondary standards were transferred to the Standards Department of the Board of Trade in pursuance of the Standards Act, 1866, and became the Board of Trade Standards. At the same time, the Standards Commission was reappointed as a Royal Commission

to consider and report upon the condition of these standards, and generally upon the subject of the secondary standards of imperial weights and measures and their verified local copies, as well as upon the system of verification and inspection of weights and measures throughout the country. The Commission met at intervals from 1867 to 1870, and as the result of their labours, presented five full and comprehensive Reports, containing many important recommendations in relation to the matters referred to them, and to the extension of the powers and duties of the Standards Department. Most of these recommendations require the sanction of the Legislature before they can be carried into effect. Their recommendations for adding to the number of secondary standards have been legalised by Her Majesty's Orders in Council, pursuant to the provisions of the Standards Act.

61. The whole series of secondary standards, including all the denominations of imperial weights and measures which can legally be used, is as follows:—

Constructed and legalised in 1824 :—

17 avoirdupois weights ...	56, 28, 14, 7, 4, 2, 1 pounds, and 10 lbs., the weight of the standard gallon of water. 8, 4, 2, 1 ounces. 8, 4, 2, 1, ½ drams.
15 troy weights	1 lb. 6, 3, 2, 1 ounces. 10, 5, 3, 2, 1 pennyweights. 12, 6, 3, 2, 1 grains.
1 measure of length ...	1 yard.

SECONDARY IMPERIAL STANDARDS.   95

10 measures of capacity ...
- 1 bushel of 8 gallons.
- 1 half-bushel of 4 gallons.
- 1 peck of 2 gallons.
- 1 gallon.
- 1 half-gallon.
- 1 quart = $\frac{1}{4}$ gallon.
- 1 pint = $\frac{1}{8}$,,
- 1 half-pint = $\frac{1}{16}$,,
- 1 gill = $\frac{1}{32}$,,
- 1 half-gill = $\frac{1}{64}$

Constructed and legalised in 1853 :—

30 decimal troy ounce weights for bullion ...
- 500, 400, 300, 200, 100, oz.
- 50, 40, 30, 20, 10 oz.
- 5, 4, 3, 2, 1 oz.
- 0·5, 0·4, 0·3, 0·2, 0·1 oz.
- 0·05, 0·04, 0·03, 0·02, 0·01 oz.
- 0·005, 0·004, 0·001, 0·002, 0·003 oz.

Constructed and legalised in 1859 :—

1 avoirdupois weight ... { 62·321 lbs. weight of cubic foot of distilled water.

3 gas-measuring standards   10, 5, 1 cubic feet.

Constructed and legalised in 1871 and 1876:

23 decimal grain avoirdupois weights .. ..
- 4,000, 2,000, 1,000 gr.
- 500, 300, 200, 100 gr.
- 50, 30, 20, 10 gr.
- 5, 3, 2, 1 gr.
- 0·5, 0·3, 0·2, 0·1 gr.
- 0·05, 0·03, 0·02, 0·01 gr.

2 Foreign postal avoirdupois weights : $\frac{1}{2}$, $\frac{2}{3}$ oz.

20 Coin weights of each coin of the realm ...
- Gold ; full weight ; 5, 2, 1, $\frac{1}{2}$ sov.
- Least current weight, 5, 2, 1, $\frac{1}{4}$ sov.
- Silver ; crown, half-crown, florin, shilling, 6, 4, 3, 2, 1d.
- B onze ; 1d., $\frac{1}{2}d$., $\frac{1}{4}d$.

WEIGHING AND MEASURING.

7 measures of length ...
{
100 feet, with sub-divisions of 10 feet.
66 feet chain, with sub-divisions of 10 links.
10 ft., divided into feet.
6 ft , divided into feet.
3 ft. or yard, divided into feet and nails or sixteenths. .
2 ft., divided into feet and inches.
1 ft., divided into inches, and 1 in. into tenths, twelfths, and sixteenths.
}

1 Liquid measure of capacity: 1 quarter gill.

2 Bottle measures ... ... { Bottle of ⅛ gallon.
Half-bottle of $\frac{1}{16}$ gallon.

4 Fluid ounce measures : 4, 2, 1, ½ oz.

16 Liquid measures of grain weights of distilled water
{
7,000, 4,000, 2,000, 1,000 gr.
500, 300, 200, 100 gr.
50, 30, 20, 10 gr.
5, 3, 2, 1 gr.
}

7 cubic inch measures : 10, 5, 2, 1, 0·5, 0·2, 0·1 cub. in.

2 Test gas-meters, passing 5 and 1 cubic feet of gas or air at one revolution.

The following total numbers of legal weights and measures of the imperial system may therefore now be used, standards of each having been constructed and verified with scientific accuracy :

Avoirdupois weights	... 63
Troy weights	... 65
Measures of length	... 7
Measures of capacity	... 45
Total	... 180

But the discontinuance of all the troy weights and their standards has been recommended by the Standards Commission.

VI.—*The Metric System.*

62. As a system of weights and measures, constructed on strictly scientific principles, the metric system may justly claim pre-eminence over all others. It was established upon the fundamental basis of the *metre*, its primary unit of length, having its length found in a determinate decimal ratio to one of the largest natural constants, that is to say, equal to the ten-millionth part of the earth's meridian-quadrant. It includes a fixed relation between the units of weight and capacity, the *kilogramme* and the *litre*, and the unit of length, the *metre*, from which both are derived; and it comprehends a uniform decimal scale of multiples and parts of these units. The more recent progress of modern science has, however, demonstrated that the actual standards of metric length, weight, and capacity do not exactly correspond with their scientific definition. The length of the meridian-quadrant is found not to be a constant, as the meridian lines vary in length according to the longitude. The latest computation by Captain Clarke of the length of the meridian passing through Paris is 10,001,472·5 metres; and that of the minimum meridian-quadrant in longitude 105° 34' is 10,000,024·5 metres. Indeed apart from the insuperable difficulties which have been found to exist in the precise determination of material standards from any natural constant, the unanimous opinion of several of the highest scientific authorities in this country has been deliberately expressed that there is no practical advantage in adopting a unit founded in

H

nature over one of an arbitrary character. The great advantage of the metric system really consists in the simplicity and uniformity of its decimal scale, and in the great convenience of this scale for all purposes of account as agreeing with the decimal system of notation, and more especially when combined with a decimal coinage which formed part of the original scheme. These undoubted advantages have proved the chief recommendations to the adoption of the metric system, first by France, and afterwards by so many other countries, and generally by scientific men. There is now every prospect of the metric system being generally adopted in all countries of the civilised world, thus greatly enhancing its value, as a common international system of weights and measures, and constituting, as it were, a universal language for expressing all quantities weighed or measured.

63. The original steps which led to the establishment of the metric system in France were taken with a view of reforming the old French system of weights and measures. These had become intolerable from their defective state and want of uniformity. In 1790, on the motion of M. Talleyrand in the National Assembly, the question of the formation of an improved system to be based upon a natural constant was referred to the French Academy of Sciences. A request was also made at the same time to the British Government that the Royal Society should act jointly with the French Academy, but no response was given to the invitation, in consequence of the distrust then entertained in this country at the progress of the revolutionary party in France. The preliminary work was

consequently entrusted to five of the most eminent members of the French Academy, Lagrange, Laplace, Borda, Monge, and Condorcet. The important Report of this Committee, which bears also the signature of a sixth member, Lalande, gave rise to the metric system. It was presented to the Academy on March 19, 1791, and is printed at length in their Memoirs. The choice of the fundamental unit of the new system lay in its derivation either from the length of the seconds-pendulum, of the earth's equator, or of the earth's meridian. The Committee rejected the length of the pendulum beating seconds as the basis of the new standard unit of length, because it involved a heterogeneous element, that of time, as well as an arbitrary element, the division of the day into 86,400 seconds. They proposed a unit of length taken from the dimensions of the earth itself, and not dependent upon any other quantity; and they did not hesitate to select the quadrant of the meridian as its basis, in preference to a quadrant of the equator, from its being a universal measure applicable to all countries, as every country was placed under one of the meridians of the earth, whilst only a few countries are under the equator. They considered also that no greater dependence could be placed upon the regularity of the equator than upon the equality or regularity of the several meridians. They recommended the ten-millionth part of the quadrant of the meridian as the definition of the new fundamental unit of length. Renouncing the ordinary subdivision of the meridian-quadrant into degrees, minutes, and seconds, they proposed a uniform decimal scale for the new system as practically the best, from its agreeing with the scale of arithmetical notation. In order that no other arbitrary principle should be introduced into the new system of weights and measures, they recommended for the basis of the unit of weight a measured quantity of distilled water, being a homogeneous substance, always to be easily found in the same degree of purity and density ; and that such quantity should be weighed in a vacuum at its temperature when passing from a solid to a liquid state.

For the practical purpose of ascertaining the length of the meridian-quadrant, they proposed to measure an arc of the meridian from Dunkirk to Barcelona, a distance of nearly $9\frac{1}{2}°$, and comprehending about $6°$ to the north and $3\frac{1}{2}°$ to the south of the mean parallel of latitude. These extreme points had also the advantage of being both at the sea level. The actual operations required were stated to be as follows :—

1. To determine the difference of latitude between Dunkirk and Barcelona.

2. To re-measure the ancient bases which had served for the measurement of a degree at the latitude of Paris, and for making the map of France.

3. To verify by new observations the series of triangles employed for measuring the meridian, and to prolong them as far as Barcelona.

4 To make observations in lat. $45°$ for determining the number of vibrations in a day, and in a vacuum at the sea level, of a simple pendulum equal in length when at the temperature of melting ice, to the ten-millionth part of the meridian-quadrant, with a view to the possibility of restoring the length of the new

standard unit, at any future time, by pendulum observations.

5. To verify carefully and by new experiments the weight in a vacuum of a given volume of distilled water, at the temperature of melting ice.

6. To draw up tables of existing measures of length, surface, and capacity, and of the different weights in use, in order to ascertain their equivalents in the measures and weights of the new system, as soon as they should be determined.

In pursuance of the recommendations of this Report, the law of March 26, 1791, was passed by the National Assembly for constructing the new system upon the proposed basis; and the Academy of Sciences was charged with the direction of the necessary operations. They entrusted the measurement of the arc of the meridian from Dunkirk to Barcelona to two of their members, Méchain and Delambre, who carried on the work during seven years, from 1791 to 1798, notwithstanding many great difficulties and dangers.

The unit of measure adopted for the actual measurement was the existing French standard of length, the toise of the Academy, better known as the *toise de Perou*, a measure of 6 French feet (*pieds du Roi*). The foot was divided into 12 inches, and each inch into 12 lines. This standard is now deposited at the Observatoire at Paris. It is a rectangular bar of polished iron 17 lines in breadth, and 4½ lines in thickness. The whole length of the bar is a little longer than a toise.

The standard length of a toise was defined by a rectangular step near each end of the bar, leaving the remaining portion at the ends half the breadth of the measuring part of the bar. It is thus an end-standard. This old French standard of length is represented in the following figure.

FIG. 26.—FRENCH STANDARD TOISE OF PERU.

The true length of the toise was taken about a line from the re-entering angles of the bar (here shown between the points of two arrow-heads a, b), when the bar was at the temperature of 13° of Réaumur's thermometer (16°·25 C., or 61°·25 F.). At this standard temperature the toise of Peru has been determined to be equal to 1·949036 metre at 0° C.; or to 76·735087 English inches, taking the equivalent of the metre to be 39·37079 inches of the Imperial Standard yard at 62° F.

This standard toise had been originally constructed as the unit for measuring an arc of the meridian in Peru, and for verifying the meridian of Paris, in 1740;

and it was substituted in 1766 for the more ancient French standard of length, the *toise du Grand Chatelet*, from which it had been originally derived. The older toise was deemed wanting in the scientific precision requisite for a standard of length. It had been constructed in 1668, and is said to have been 5 lines shorter than the toise measure then ordinarily used, for which no authoritative standard could be found; and to have been actually derived from the width of the inner gate of the entrance to the Louvre, which, according to the original plan, was made 12 feet wide. One half of this width was taken for the length of the standard toise.

The measures actually used for the survey operations are known as the *règles de Borda*, though it is now known that they were constructed under the immediate superintendence and direction of the celebrated Lavoisier, who took the most active part in the operations for establishing the metric system, until his death on the scaffold in 1794. These measures were four in number, each consisting of a bar of platinum two toises, or 12 French feet in length, about $\frac{1}{2}$ inch broad, and $\frac{1}{12}$ inch thick. Each platinum bar was fixed at one end only to a bar of brass about $11\frac{1}{2}$ feet long, the other end of the platinum bar being free and extending about 6 inches beyond the corresponding end of the brass bar. The object of this second bar was that it should form, together with the first bar, a metallic thermometer, indicating the temperature of the two bars by their difference of dilatation, which could be measured by a fine vernier. The four measuring bars were accurately verified, and found, when placed together, end to end, not sensibly to differ from eight times the length of the toise of Peru at the temperature of 12 ·5 C.

The base for the measurement of the northern portion of the work was measured at Melun, and found to be 6075·90 toises. The base for the southern portion was measured at Perpignan, and found to be 6006·25 toises.

64. The final results of all the operations for determining the new metric unit of length, were stated by the Commission in their Report, dated April 30, 1799. They found :—

1. That the length of the arc of the meridian comprehended between Dunkirk and Barcelona, was $9°6738$ (or 9° 40′ 45″), and measured 551,584·72 toises.

2. Assuming, from the previous measurements in France and Peru, that the mean ellipticity of the earth was $\frac{1}{334}$, they computed the length of the meridian quadrant to be 5,130,740 toises.

3. That the length of the new unit of length, the ten-millionth part of the meridian-quadrant, was equal to 0·5130740740 toise, or 3 feet and 11·296 lines; being 443·296 lines of the toise of Peru (which contained 864 lines), at its standard temperature of 16°·25 C. In terms of the new standard unit, the toise of Peru was equal to 1·949036591 metre.

4. That the length of the pendulum at the temperature of melting ice, beating seconds in a vacuum at the sea level at Paris, was equal to 0·99385 metre.

65. The actual construction of the new standard measure of length had been entrusted to the mechanician Lenoir. As a preliminary proceeding, he made

THE METRIC SYSTEM. 105

four end-standard metres of brass, differing in length very slightly from each other, and each about equal to 443·242 lines of the toise of Peru. This was the computed length of one ten-millionth part of the meridian-quadrant, as deduced from the previous measurements of an arc of the meridian in France made in 1740.

Of these four metres, No. 2 was found after several comparisons to be nearest to the required length. It was accordingly selected as the provisional standard metre.

For obtaining the definitive standard metre, two standard metres of platinum, and twelve metres of iron, were constructed by Lenoir, who had improved his comparing apparatus so as to show differences of 0·001 line. The Commission were not satisfied with making numerous comparisons of these metres and the provisional metre of brass among themselves, but they also compared them repeatedly with the four *règles de Borda* and a new supplementary measure of above 45 lines, so as to determine not only their relative and absolute length, but also the rates of expansion of the three metals of which they were composed. The rates of expansion definitively adopted by the Commission, from observations made by Borda between $0°$ and $32°$ C., were as follows :—

In a metre.
Coefficient of linear expansion of platinum for $1°$ C. $= 0·00000856$, or $0·0031$ mm.
,, ,, brass ,, $= 0·00001783$, or $0·0092$,,
,, ,, iron ,, $= 0·00001156$, or $0·0063$,,

The comparisons and corrections of the several metres were continued until no difference amounting to 0·0000001 toise, or 0·001 millimetre, could be found at the temperature of melting ice, either in their desired absolute length of 443·296 lines of the toise of Peru or in relation to each other. They were consequently all determined to be perfectly exact. One of the platinum metres, subsequently known as the *metre des Archives*, from its place of deposit, was reserved as the new prototype measure of length ; the other was kept at the Observatoire at Paris, as its accessible representative. The twelve iron standard measures were distributed amongst the several countries represented at the Commission.

The primary *metre des Archives* is a rectangular platinum bar, bearing no mark or inscription. Its breadth is 25 mm. (0·984 in.), its height 3·5 mm. (0·138 in.). Its ends are planes perpendicular to its axis of length, and the straight line between them in this axis denotes the true length of the metre at 0° C., or the temperature of melting ice. It thus constitutes what is termed a *mètre-à-bouts* or end-standard metre.

66. The unit of metric weight was defined to be the weight in a vacuum of a cubic decimetre of distilled water at its maximum density, or the temperature of 4° C. Distilled water was selected as the best material in nature for thus determining the unit of weight, from its being obtainable everywhere and at all hours in the greatest purity, its being perfectly homogeneous, and its density being invariable at any given temperature. It was required first accurately to ascertain the weight of this volume of water, and then to construct a metallic standard of equivalent weight. There were two modes of proceeding for determining the volume of water to be weighed ; one, by measuring

the internal capacity of a vessel to contain this volume of water; the other, by measuring externally a solid or hollow body, in order to ascertain the weight of the volume of water displaced by it. This last method was preferred, it being considered that the accurate external measurement of a metallic body was much less difficult than that of the internal capacity of a metallic vessel; and it was determined that the best form of this body was a cylinder of a height equal to the diameter of the base, this form being capable of being made and measured with the greatest precision.

It was not thought requisite that the cylinder should be of the specified volume of a cubic decimetre, but only of the most convenient size for arriving at the desired result by computation. The cylinder actually used was made of brass, and hollow, being only so much heavier than the same bulk of water as to enable it to sink by its own weight when plunged in water. It was intended to be 2·435 decimetres (about $9\frac{1}{2}$ inches) in diameter and height.

To facilitate the accurate measurement of the cylinder, 12 radial lines or 6 diameters are said to have been drawn on its base plane, dividing it into twelve equal parts; and corresponding lines were drawn on its upper plane. The ends of these two series of lines at the circumference were joined by vertical lines on the cylinder, thus dividing it vertically into twelve equal parts. Circular lines were also traced on the two plane surfaces at about 11 mm. from the circumference, and at half and two-thirds of the radius from the centre; and eight horizontal lines were drawn around the cylinder at the following distances from the base:—13, 35, 67, 95, 148·5, 176·5, 208·5, 230·5 millimetres. The height of the cylinder was determined from the ascertained mean distance of

FIG. 27.—CYLINDER FOR DETERMINING WEIGHT OF CUBIC DECIMETRE OF WATER (¼th size).

the corresponding 37 points of intersection of the lines on the upper and lower surfaces, including the centres. The diameter of the cylinder was determined from the ascertained mean length of the 48 diameters, included between the corresponding points of intersection on its cylindrical portion. It should, however, be stated that no lines are now visible on the surface of the cylinder, which is deposited at the Observatoire at Paris.

The measurement was effected by means of an apparatus specially constructed for the purpose by Fortin, and it indicated minute differences of length of $\frac{1}{1000}$ line, or $\frac{1}{1100}$ mm. The standard measures used for determining the absolute length measured were 16 brass measures, specially constructed for the purpose, each very nearly equivalent to the height of the cylinder, and 16 other measures, each nearly equivalent to its diameter. The length of each of these two series of measures in relation to each other was ascertained by numerous observations with the new apparatus; and the total length of each set of 16 measures in relation to the new standard unit was obtained by comparing the sum of their length with Borda's *règle* of 2 toises, No. 1 (to which they very nearly corresponded in length), by means of the *comparateur* used for the comparison of these large measuring rules.

The final result of the measuring operations was that the mean height of the cylinder was determined to be 2·437672 decimetres, and its mean diameter 2·428368 decimetres, at the temperature of 17°·6 C. According to Borda's determination of the coefficient of the linear expansion of brass, the volume of the cylinder was determined by computation to be nearly 11·28 cubic decimetres, when at the temperature of melting ice.

For ascertaining the weight of water displaced by this cylinder, a series of brass weights was specially constructed, consisting of a unit or provisional kilogram, made as nearly as possible of the estimated weight of a cubic decimetre of water, together with 11 exact copies and smaller weights in decimal subdivision down to the millionth part, all carefully verified and deemed to be accurate within less than half of one millionth-part

The mean weight of the cylinder in ordinary air was taken, no reduction to a vacuum being deemed requisite, as the weights used were of similar metal to the cylinder, the interior of which communicated with the external air. For this purpose a metallic tube, 1·285 mm. in diameter, was screwed to the top of the cylinder, its end being out of the water when the cylinder was immersed. The top of the cylinder was 43 mm. from the surface of the water during the weighings, and the volume of the tube immersed was therefore 55·77 cubic mm. Taking the volume of the cylinder to be 11·28 cubic decimetres, the volume of the metallic part of the cylinder was computed to be 1·506 cub. decim., and of the hollow part filled with air 9·774 cub. decim. During the weighings the cylinder was surrounded with ice, but the temperature of the water was never below $0°·2$ C. and the mean temperature was $0°·3$. The final results of the weighings were declared to be as follows:—

Weight of the cylinder in air, in terms of the
unit employed ... ... = 11·4660055
Its mean weight in distilled water, after deducting the weight of air in the cylinder, and of
the air displaced by the weights used . = 0·1967668
Hence weight of the volume of distilled water
equal to the volume of the cylinder ... = 11·2692387.

At the time when the metric system was originated the French standards of weights were the series known as the *pile de Charlemagne*, the unit being the *livre poids de marc* of 16 *onces*, and double the *poids de marc*. The metric equivalent of the *poids de marc* was subsequently determined to be 244·753 grammes. The *once* was divided into 8 *gros* (or drachms), and the *gros* into 72 *grains*. The old French *livre* of 9216 French *grains* was therefore equal to 489·506 grammes or to 7554 English troy grains. The French grain *poids de marc* was equal to 0·818 English troy grain. In determining the new unit of metric weight, it was necessary to ascertain the actual value, in terms of the existing system of the *livre* and its subdivisions, of the provisional weights used; and from accurately comparing them with the old standard, it was deduced from the ascertained weight of the measured cylinder, that the weight of a cube decimetre of distilled water at its maximum density, or at 4° C., which was 0·9992072 of the provisional kilogram, was equal to 18827·15 grains of the *poids de marc*. This, accordingly, was definitively adopted as the true weight of the kilogram, the new unit of metric weight.

The determination by the French Commission of the weight of a cubic decimetre of water at its maximum density differs somewhat from later authoritative determinations made in England and other countries, as may be seen from the following tabular statement :—

Date.	Country.	Observer.	Weight of cubic decimetre of distilled water at 4° C.
			Grammes.
1795	France ...	Lefevre-Gineau ...	1000·000
1797, 1821	England ...	Shuckburgh and Kater	1000·480
1825	Sweden ...	Berzelius, Svanberg, and Akermann ...	1000·296
1830	Austria ...	Stampfer	999·653
1841	Russia ...	Kupffer	999·989
		Mean	1000·084

But the latest and most carefully executed determination by Kupffer agrees so nearly with the French determination, that the actual weight of the primary kilogram may be taken as almost identical with its theoretical definition, and sufficiently accurate for all practical purposes.

67. From the provisional brass kilogram, with its error thus ascertained by the French Commission, two new standard kilograms were constructed by Fortin, one of platinum, the other of brass, and each was determined, after numerous comparisons and the requisite corrections, to be of the true weight when weighed in a vacuum. The platinum weight was constituted the primary metric standard kilogram, and is known as the *kilogramme des Archives*. Its form is that of a cylinder of about 39·4 millimetres in diameter, and 39·7 millimetres high, having its edges slightly rounded, being similar to that of the English platinum kilogram shown of the actual size in Fig. 30. The density of the *kilogramme des Archives* has never been precisely determined, as it has been deemed hazardous to weigh it in water, from a fear of its not

being entirely free from the arsenic used in preparing the platinum, and of dissolving this arsenic, and thus diminishing the weight of the kilogram. Prof. Miller has computed the volume of the *kilogramme des Archives*, when at its normal temperature of 0° C., to be equal to the volume of 48·665 grammes of water at its maximum density, as determined by its cubic measurement, and consequently its density to be 20·5487. Other computations, however, differ slightly from this determination.

The brass kilogram was intended as the commercial standard, for regulating all ordinary metric weights in air, and was deposited at the Ministère de l'Intérieure, Paris. One uniform shape is adopted in France for all brass kilograms. They are made in the form of a cylinder surmounted with a knob. The height of the cylinder is equal to its diameter, and the height and diameter of the knob are equal to one half those of the cylinder. Like the platinum *kilogramme des Archives*, the brass standard kilogram was never weighed in water, and its volume has been computed from its cubic measurement to be equal to that of 124·590 grammes of water at its maximum density, thus making its density 8·206. In our English standard air, $t = 62°$ F. $b = 30$ in., the platinum standard kilogram will thus displace 59·25 milligrams of air, and the brass kilogram 151·75 mgr.; the apparent weight in air of the brass kilogram is consequently about 92 mgr. less than that of the platinum standard. This brass kilogram was assumed by the French Commission to be 88·5 mgr. lighter than the platinum standard, when weighed in ordinary air.

The primary platinum metre and kilogram were presented by the Commission, on June 22, 1799, to the Corps Legislatif at Paris, and were legally constituted as the standards of length and weight of the new metric system throughout France by the law of December 9, 1799. They were deposited at the Palais des Archives.

A platinum copy of each of the primary metric standards of the metre and kilogram was constructed at the same time, and deposited at the Paris Observatory. These standards, known as the *mètre de l'Observatoire*, and the *kilogramme de l'Observatoire*, were considered as next in authority to the primary standards.

68. The unit of capacity of the metric system, the *litre*, represents theoretically the measure of volume of a cubic decimetre, or the cubic contents of a metallic vessel of this capacity when the metal is at the temperature of melting ice. But practically, there is no material primary standard litre, and the legal measure of the litre is determined from the kilogram ; that is to say, the litre actually is a measure containing a kilogram weight of distilled water at its maximum density. Such a measure can only be verified by computation, as the vessel itself must be taken at a different temperature from the water contained in it, the vessel at 0° C., the water at 4° C. Authoritative tables are therefore prepared for ascertaining the allowance to be made in every case for differences of temperature from the normal temperature, as well as for the difference of weight of air displaced by the metallic weight and the larger volume of water.

For metric measures of surface, the *are*, equal to

THE METRIC SYSTEM.

100 square metres, is the unit; and for solid measures, more particularly for measuring wood, the *stere*, or cubic metre, is the unit.

69. The number and denominations of the metric weights and measures actually used in France and other countries, for which specific standards are provided, together with their names, are as follows: they include the double and the half of each decimal unit, with a duplicate unit to make up the number 9 units :—

6 metric measures of length ...	Double metre. Metre, divided into tenths or decimetres, &c. Demi-metre ,, ,, Double decimetre, divided into centimetres and millimetres. Decimetre, divided into centimetres and millimetres. (For land) Chain of double dekametre, or 20 metres, divided into metres, and links of 2 decimetres.
30 metric weights	20, 10, 5, 2, 1, 1 kilograms. 500, 200, 100, 100 grammes (hectograms). 50, 20, 10, 10 ,, (dekagrams). 5, 2, 1, 1 ,, 0·5, 0·2, 0·1, 0·1 gramme (decigrams). 0·05, 0·02, 0·01, 0·01 gramme (centigrams). 0·005, 0·002, 0·001, 0·001 gramme (milligrams).
13 metric measures of capacity...	Hectolitre, . or 100 litres. Demi-hectolitre, ,, 50 ,, Double dekalitre, ,, 20 ,, Dekalitre, ,, 10 ,, Demi-dekalitre, ,, 5 ,, Double litre, ,, 2 ,, Litre, ,, 1 litre. Demi-litre, ,, 0·5 ,, Double decilitre, ., 0·2 ,, Decilitre, ,, 0·1 ,, Demi-decilitre, ,, 0·05 ,, Double centilitre, ,, 0·02 ,, Centilitre, ,, 0·01 ,,

Total number of metric weights and measures used in France and other countries, 49.

For dry commodities, the demi-dekalitre is the smallest measure used. The litre being equal to a cubic decimetre, or 1,000 cubic centimetres, in volume, is also equal to 1,000 grammes weight of distilled water at its maximum density; consequently the

Demi-litre	= 500	cubic centimetres,	or grammes weight	of water.
Double decilitre	= 200	,,	,,	,,
Decilitre	= 100	,,	,,	,,
Demi-decilitre	= 50	,,	,,	,,
Double centilitre	= 20	,,	,,	,,
Centilitre	= 10	,,	,,	,,

There are also graduated measures of 5, 2, and 1 cubic centimetres or grammes weight of water.

70. The earliest recognition by the British Parliament of the metric system thus established in France took place soon after the close of the French war. On March 15 1816, Mr. Davies Gilbert brought forward a motion in the House of Commons, which was carried, for comparing the imperial standard yard with the French standard metre. The Government entrusted the necessary operations to the Royal Society, who obtained for the purpose two platinum metres from Paris. These had been verified by M. Arago, by comparison with the French standard. One was an end-standard, like the "Metre des Archives," but was nearly twice as thick, being 7·3 millimetres in thickness. On one plane surface the word "METRE" is engraved, and on the other "FORTIN A PARIS," and "Royal Society, 44." This end-standard was determined to be exactly the

length of a metre at the temperature of melting ice. The other was a line-standard, the bar being nearly equal in width, but only 5·3 millimetres thick, and it is about 4 centimetres longer. On the upper surface is engraved "Royal Society, 45," and transverse lines, so fine as hardly to be seen with the naked eye, and terminating in arrow-heads, are cut about 2 centimetres from each end for defining the length of the metre, as shown in the following figure :—

FIG. 28.—DEFINING LINES OF PLATINUM LINE-STANDARD METRE OF ROYAL SOCIETY.

The length of a metre is to be taken between the two transverse lines at the mid-width of the bar, and it has been determined to be less than a metre by 0·01759 millimetre, at the standard temperature of melting ice.

71. On being brought to this country, the two platinum metres were carefully compared by Captain Kater with the length of 39·4 inches on the Shuckburgh scale, considered by him to be the British scientific standard of length. Full details of the comparisons made with Captain Kater's microscopical comparing apparatus are given in Phil. Trans. 1818. It was required to determine the length of the

platinum metre at its standard temperature of 32° Fahr. in terms of the brass standard yard of 36 inches at its standard temperature of 62° Fahr. Allowance was made for the different rates of expansion of the two metals, the coefficient of expansion of the platinum being taken to be 0·00000476 for 1° Fahr., as determined by Borda, and that of brass 0·0000101, as found by Kater's experiments. The length of the metre at 32° Fahr. was thus determined from the *mètre-à-bouts* to be 39·37086 inches of the Shuckburgh scale at 62° Fahr., and from the *mètre-à-traits* 39·37081 inches, after allowing for its error = 0·00069 inch. The mean length of the metre was therefore 37·37084 inches of the Shuckburgh scale, and as this scale had been found 0·00005 inch longer than the Parliamentary standard, the true length of the metre was finally determined by Captain Kater to be 39·37079 British inches.

Ever since this period, this authoritative equivalent of the metre in imperial measure has been recognised as the true equivalent, and it received the sanction of Parliament, in the Act of 1864, for legalising contracts made in this country in terms of the metric system. It is, however, to be observed that it is the *scientific* equivalent of the metre in imperial measure, when each standard is taken at its own normal temperature. For all *commercial* purposes, on the other hand, the measure of a metre is always used at ordinary temperatures just as a yard measure is used, and the comparison of the two should therefore be more properly made at the same average temperature of 62° F. At such temperature a brass metre is equal to 39·382 inches, and this length is to be taken as the true commercial equivalent of the metre in British measure. Of course, this difference between the equivalent in imperial measure of the metre at its legal and at its ordinary temperature, amounting only to $\frac{3}{1000}$ inch, is perfectly immaterial in commercial measurements of small quantities, and the metre may safely be estimated as equal to $39\frac{3}{8}$ of our inches, and the decimetre at 3·94 inches, as shown in the following figure:—

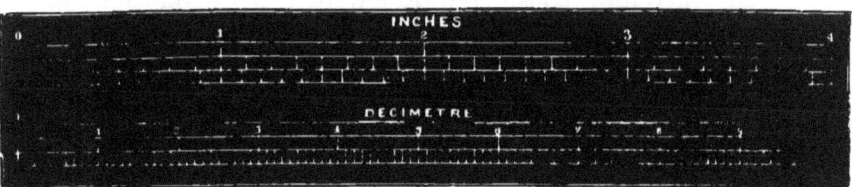

FIG. 29.—DECIMETRE, AND NEARLY EQUIVALENT LENGTH OF 4 IMPERIAL INCHES.

72. No satisfactory comparison of the primary kilogram with our unit of imperial weight was made until the year 1844, after the construction of the new Imperial standard pound, under the authority of the Standards Commission. The comparison of the standard units of weight of the two countries was then undertaken by Professor Miller, at the request of the Commission. He found that previous determinations of the weight of the kilogram varied amongst themselves from a minimum of 15432·295 gr. to a maximum of 15438·355 grains. Under these circumstances, he proceeded to Paris in the autumn of 1844, and obtained permission from the French Government to compare the kilogramme des Archives

with our English weights. For the comparison he took with him the Parliamentary platinum copies Nos. 1 and 2 of the standard pound, and two auxiliary platinum weights together equal to about 1432·35 grains. The mean result of 60 comparisons was to find the kilogramme des Archives equal to 15432·34813 grains. But Professor Miller was not satisfied with this result, as one of the auxiliary weights was found to contain a small cavity filled with some hygroscopic substance, which rendered its weight slightly variable. He therefore considered it requisite to make further comparisons directly with the English standard pound.

For this purpose, a platinum kilogram, constructed by Gambey, was procured at Paris by Professor Miller, and was accurately compared by him with the kilogramme des Archives. This platinum kilogram, designated as ℭ by Professor Miller, is similar in form to the prototype, but is a little smaller, in consequence of the somewhat greater density of the platinum of which it is composed. Its density was determined by hydrostatic weighings to be 21·13791. From the mean of 100 direct comparisons with the kilogramme des Archives, ℭ was found to be lighter in a vacuum than the French standard by 1·56 mgr. (0·02412 gr.). For ascertaining the weight of ℭ in terms of the new imperial standard pound, Professor Miller subsequently compared this kilogram with the imperial standard pound, together with each of its Parliamentary copies successively, and one of four auxiliary platinum weights each of 1432·324 grains, made up for the purpose, and accurately verified in terms of the imperial standard, by means of supplementary platinum weights. The mean result of 166 direct comparisons of ℭ was to find its value = 15432·32462 grains. The kilogramme des Archives was consequently determined to be equal in a vacuum to 15432·34874 imperial grains, or 2·20462125 standard platinum lbs.; and the imperial standard pound equal to 453·5926525 metric grammes. These equivalents have since been generally accepted, and were legalised in this country by the Metric Act, 1864.

73. The platinum kilogram ℭ has since been deposited in the Standards Department, together with

FIG. 30.—ENGLISH PLATINUM STANDARD KILOGRAM ℭ (actual size).

a second kilogram, of gilt gun-metal, also made under Professor Miller's directions, and intended as a standard for the adjustment of commercial metric weights, like the French *kilogramme laiton* deposited at the Ministère de l'Intérieure at Paris. This gilt gun-metal kilogram was constructed by Oertling and has been denoted as 𝔅 by Professor Miller. Its form is spherical with a knob. Its density is 8·3291. The mean result of 24 comparisons with ℭ showed that in a vacuum the weight of 𝕽 was 1·47 mgr. less than ℭ, and 3·04 mgr. less than the kilogramme des Archives. In standard air ($t. = 18°\cdot7$ C., $b. = 755\cdot64$ mm.) 𝕽 displaced 143·92 mgr. and the kilogramme des Archives

FIG. 31.—ENGLISH GILT GUN-METAL STANDARD KILOGRAM 𝕽.

58·36 mgr. 𝕽 was thus found to be 88·6 mgr. lighter in air than the French platinum prototype, and only 0·06 mgr. lighter than the French commercial brass standard kilogram.

74. In the metric system, the nomenclature of the

several denominations of metric weights and measures, so far as regards the principal units and their decimal multiples and parts, was adopted upon a simple and uniform rule, so as to enable each name to denote the position of the measure or weight in the decimal scale. It is formed by prefixing to each unit the terms *deka, hekto, kilo,* and *myria,* derived from the Greek words for 10, 100, 1,000, 10,000, as applicable to the series of decimal multiples, and the terms *deci, centi,* and *milli,* derived from the Latin, for the decimal parts. The whole series of decimal units is shown in the following table, showing also their imperial equivalents at the scientific values, as now determined, that is to say, assigning to each system its normal temperature and condition :—

METRIC MEASURES OF LENGTH.

		Imperial Equivalents.
Millimetre	= 0·001 metre	0·03937 inch.
Centimetre	= 0·01 ,,	0·39370 ,,
Decimetre	= 0·1 ,,	3·93708 inches.
METRE	= 1 ,,	39·37079 ,,
Dekametre	= 10 metres	10·93633 yards.
Hektometre	= 100 ,,	109·36330 ,,
Kilometre	= 1,000 ,,	1093·63305 ,,
Myriametre	= 10,000 ,,	6·21382 miles.

METRIC WEIGHTS.

		Imperial Equivalents.
Milligram	= 0·00110 gramme	0·01543 grain.
Centigram	= 0·01 ,,	0·15432 ,,
Decigram	= 0·1 ,,	1·54323 ,,
GRAMME	= 1 ,,	15·43234 grains.
Dekagram	= 10 grammes	154·32348 ,,
Hektogram	= 100 ,,	1543·23487 ,,
Kilogram	= 1,000 ,,	15432·34874 ,,
Myriagram	= 10,000 ,,	22·04621 lbs. av.
Quintal	= 100,000 ,,	1·96841 cwt.
Millier, or ton	= 1,000,000 ,,	0·9482 ton.

Metric Measures of Capacity.

		Imperial Equivalents
Centilitre	= 0·01 litre	0·07045 gill.
Decilitre	= 0·1 ,,	0·17614 pint.
Litre	= 1 ,,	0·88072 quart.
Dekalitre	= 10 litres	2·20180 gallons.
Hektolitre	= 100 ,,	0·27522 bushel.
Kilolitre	= 1000 ,,	3·44030 quarters.

Metric Measures of Surface, or Square Measures.

		Imperial Equivalents.
Centiare	= 1 square metre	1·19671 sq. yard.
Are	= 100 square metres	119·67163 sq. yards.
Hectare	= 10,000 ,,	2·47255 acres.

Metric Measures of Solidity, or Cubic Measures.

		Imperial Equivalents.
Cubic centimetre	= 0·001 C.D.	0·06108 cub. inch.
Cubic decimetre	= 1,000 C.C.	0·03534 cub. foot.
Cubic metre	= 1,000 C.D.	1·30914 cub. yard.

75. Although the metric system was originally established in France as the legal system of weights and measures in 1799, it was not until more stringent provisions of law for enforcing its exclusive use were passed in 1837, that metric weights and measures began to be generally used throughout the country. Since that period, it has been gradually adopted in most of the countries of the civilised world, and there is now every prospect of its finally becoming universally in use, as an international system of weights and measures. Its introduction into this country, permissively, so far as regards the use of metric weights and measures for all purposes of trade, and exclusively for all international purposes, has been unanimously recommended by the late Standards

Commission, and is now only awaiting the decision of the legislature.

76. In 1870, an International Metric Commission was appointed to meet at Paris, formed of fifty delegates from twenty-nine of the principal countries of the world, including this country, and comprising many persons of the highest scientific eminence. The object of the Commission was the construction and verification, with all the best appliances of modern science, of a new and uniform series of metric standards of the metre and kilogram, for all countries who have adopted or contemplate the adoption of the metric system.

Many important resolutions were passed by the Commission at their first general meeting in 1872. It was agreed that the new international standards should be based upon the existing primary metric standards, the metre and kilogramme des Archives; and that no alteration of these prototypes was desirable or expedient, although they might not precisely agree with their scientific definition. For the material of the new standards, both of length and of weight, platinum, with ten per cent. of iridium, was selected, an alloy found to possess more than any other known substance the essential requisites for standards of durability and invariability, as well as other special properties desirable in standards of the greatest precision. A new sectional form of bar has been fixed upon for the international metre, which is to be a *mètre-à-traits*, or line standard, and this form appears to meet all the requirements, and to combine many advantages of a geometrical, mechanical, thermical, and economical character. It gives the greatest strength to the bar, with the least quantity of metal. It affords great rigidity in combination with the high elasticity of the metal. It enables the bar readily to take a uniform temperature, and allows the defining lines, as well as any number of subdivisions, to be marked on a plane in the neutral axis of the bar. The Commission resolved to furnish also *mètres-a-bouts*, or end-standards, to those countries who may wish for them, and that the new form of bar, with a very slight modification, should also be applicable to them.

The form of the transverse section of the new *mètre-à-traits* is shown in Fig. 32, and of the *mètre-à-bouts* in Fig. 33, that of the existing Standard *mètre des Archives* being shown in Fig. 34, all of the actual size.

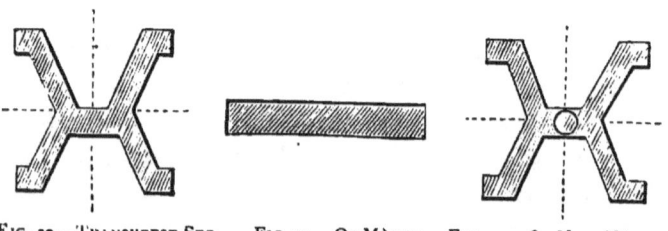

FIG. 32.—TRANSVERSE SECTION OF NEW MÈTRE-À-TRAITS.   FIG. 34.—OF MÈTRE DES ARCHIVES.   FIG. 33.—OF NEW MÈTRE-À-BOUTS.

The bar for the new *mètre-à-traits* was determined to be 102 centimetres long, and 2 centimetres square in section, reduced by chiselling and drawing it longitudinally to the specified form. The metal thus reduced was to be 3 millimetres in thickness, the horizontal breadth at the upper and lower surfaces being

4 millimetres. In the new *mètre-à-bouts*, the sectional figure is to be symmetrical, and the defining points of the metre to be the centres of the spherical ends, the radius of curvature being 1 metre. From the great rigidity of this form of bar, combined with the high elasticity of the platinum iridium, it will compare very advantageously with the mètre des Archives, the rigidity being as 25·9 to 1, whilst the superficial area of the transverse sections of the two bars will be respectively 150 and 100 square millimetres.

The form of the new international kilogram was determined to be the same as that of the kilogramme des Archives, already described, a cylinder of equal diameter and height, with the edges slightly rounded.

At the general meeting of the Commission in 1872, the actual construction of the new metric standards was entrusted to the French Section of the Commission resident in Paris, with the concurrence of a Permanent Committee composed of twelve members of the Commission, elected for the purpose, and representing twelve of the principal countries interested. At the present time (the close of the year 1876), great progress has been made in the construction of the new standards, and it is expected that they will be ready at an early period for commencing the comparisons with the standards of the Archives.

The definitive comparisons of the new standards, both with the old standards of the Archives and with each other, were entrusted to the Permanent Committee. In order that the Committee might be furnished with the requisite means for carrying out their operations, and that the new metre prototypes might be authoritatively recognised by the several countries interested as their common property, and provisions be agreed upon for their custody and future use, a diplomatic Conference was convoked and held in Paris in the spring of 1875, when a convention was entered into by the great majority of the plenipotentiaries then assembled.

77. Under the terms of this convention, an International Metric Bureau has been established at Paris, and provision made for the necessary buildings, apparatus, and administrative officers, placed under the superintendence of the Permanent Committee, which has been newly constituted as the International Metric Committee (*Comité International des Poids et Mesures*). The expenses of the establishment and annual maintenance of the Bureau are to be defrayed by rateable contributions from the several countries who are parties to the convention. All the definitive comparisons of their new metric standards for the contracting countries are to be carried out in this new institution, where the future international metric prototypes are to be deposited under specific regulations for their custody, preservation, and use. The institution, with its equipments, is also to be made available generally for metrological purposes and for comparisons of standard weights and measures of precision, not only for the several governments, but for public bodies and even individual men of science.

Another effect of the convention is to abrogate the International Metric Commission, and to transfer its duties and functions to the new Metric Committee, reserving only the duties entrusted to the French section of the construction and first comparisons of the new metric standards. The Pavillon Breteuil in the park of St. Cloud with its dependencies has been given up by the French government as the site of the new metric institution, and the Committee have announced that they hope to have all the internal arrangements completed so as to begin their scientific operations in the course of the year 1877.

It is to be regretted that this country takes no part in this new scientific institution. Her Majesty's government have declared that they cannot recommend to Parliament any expenditure connected with the metric system, which is not legalised in this country, nor in support of a permanent institution established in a foreign country for its encouragement. They have consequently declined to take part in the convention, or to contribute towards the expenses of the new Metric Bureau, and they have directed the Warden of the Standards, who, as one of the delegates from this country to the International Metric Commission, had been elected a member of the Permanent Committee, to decline being appointed a member of the new International Metric Committee, or to take part in the direction of the new International Metric Bureau. They have also requested the French government to deliver over the new international metric standards required for the country at the earliest period after their construction and verification by the French section of the Commission. It will however be always open to this country under the terms of the convention either at any time to take

K

part in and contribute to the new International Metric Bureau, or to have the new metric standards, which have been so verified by the French section, compared at this institution in the same way as the new metric standards of contracting countries.

VII. *Weighing and Measuring Instruments, and their Scientific Use.*

78. The instrument universally used for weighing is the balance, with its various modifications. It serves to determine the weight of bodies by comparison with a body of known weight, such as a standard weight. The simplest form of balance is a beam made to vibrate upon a centre of motion, with pans, or other contrivances for supporting the bodies weighed, hanging from the extremities of the two arms of the balance. From the depression of either pan, the excess of weight of the body placed in it is determined.

Balances are of two kinds :—1. Ordinary balances, or scale beams, with equal arms, which have the beam suspended by the middle. If an equal-armed balance is properly adjusted, so that the beam is exactly horizontal when the pans are empty, the beam will be in equilibrium, and the balance will also be in equilibrium, that is to say, the beam will rest in a horizontal position, after equal weights have been placed in the pans. 2. Balances with unequal arms, in which the beam vibrates upon the centre of motion placed more or less near one of the extremities. These two classes comprehend all balances of precision or scientific balances. Spring balances are also used, in which the body weighed pulls down a spring, to which a pointer is attached and moves over a graduated scale, thus indicating the weight. But these balances are not balances of precision.

In both of the above-mentioned kinds of balance, the beams are levers of the first order, the fulcrum upon which the beam vibrates being placed between the power and the weight, that is to say, between the points of the beam which carry the bodies compared. On the principle of the lever, the power of any weight to move a balance is proportionately greater according as the part of the beam which supports that weight is more distant from the fulcrum or centre of motion of the balance. Hence it follows that the power of the weight to move a balance beam, is in a ratio compounded of the weight itself and of the distance of its point of suspension from the centre of motion of the balance. A multiplying or proportionate balance may consequently be constructed for determining the weight of a body placed in the pan suspended from the shorter arm of the beam, by its being found equal to a multiple of a unit weight in the pan suspended from the longer arm of the beam, and usually termed the weight pan. For this purpose, if the beam be divided into, say, three equal parts, and the centre of motion be placed at the first division, or at one third of the whole length of the beam, it will be seen that ·1lb. placed in the weight pan will form an equipoise with 2lbs. in the other pan, and so on. This principle is greatly extended in larger weighing machines by lengthening the longer arm, and by the use of compound levers, so that 1lb. can be made to form an equipoise with 100lbs. or more.

79. The ancient Roman balance is perhaps the earliest form of a well-constructed multiplying balance, and corresponds with our modern steelyard. It has been remarked by Sir Gardiner Wilkinson, that no instance has been found of the existence of the steelyard before the Roman era. But the principle of its construction was in use amongst the ancient Egyptians, who ascertained the weight of articles suspended from different parts of a scale beam by means of a heavy determinate weight placed in one scale. The Roman balance consists of a determinate weight attached to the longer arm of the beam, and made to traverse along a number of divisions marked upon it. The multiplied power of the traversing weight when resting on the several sub-divisions, as they increase in distance from the centre of motion, is indicated by corresponding figures upon the graduated beam.

The following figure (taken by permission from the "Imperial Journal of Art," vol. i. p. 85) represents an ancient Roman balance of an elegant form, found at Pompeii, and in use A.D. 77. It is described as having the graduated divisions on the longer arm of the beam marked with Roman numerals from X. to XXXX. (probably Roman pounds), and with a V. on the half of each decimal series, the smaller sub-divisions being also marked. The inscription on the shorter arm of the beam (shown in a separate and enlarged figure) denotes its having been proved at the Capitol in the eighth year of Vespasian Emperor Augustus, and in

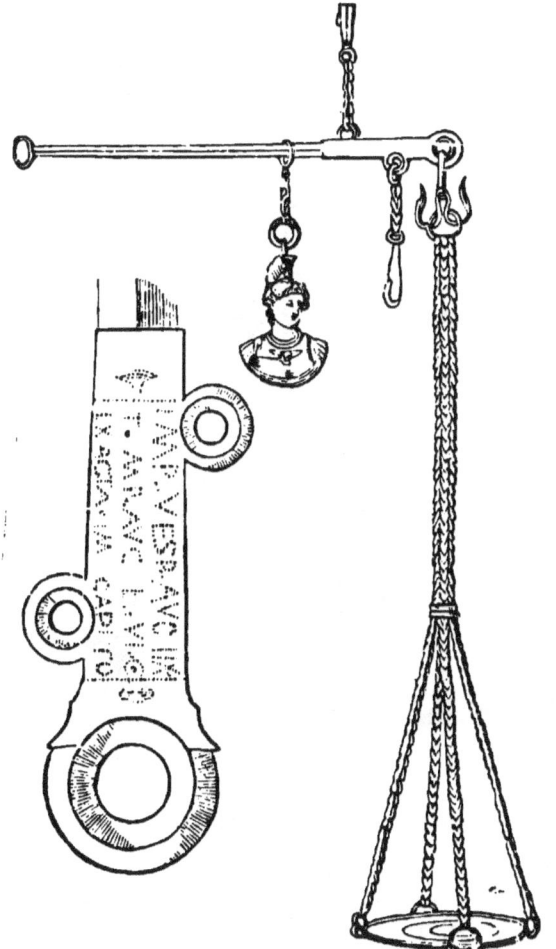

FIG. 35.—ANCIENT ROMAN BALANCE OF VESPASIAN.

the sixth consulate of Titus Emperor Augustus 'his

son. This steelyard is consequently a duly verified standard weighing machine.

There is undoubted evidence of the use of equal-armed balances by the ancient Egyptians. In a very beautiful Egyptian papyrus of Hunnafer, superintendent of the cattle of Sati I., 19th Dynasty, about 1350 B.C., known as the "Ritual of the dead," there

FIG. 30.—ANCIENT EGYPTIAN EQUAL-ARMED BALANCE.

is a representation of a well-constructed equal-armed balance, as shown in Figure 36, taken by permission from a photograph of this papyrus, one of a series of photographs from the collections at the British Museum, made by S. Thompson, and published by Marsell and Co., 2, Percy Street.

In the original papyrus, the middle and both ends

of the beam, as well as the lower part of the column are coloured to represent polished brass; the other parts of the balance are dark, as if of bronze. It may be noticed that there are box ends to the beam. The colours of the papyrus are fresh and bright, as when first painted. The soul of the deceased is shown as being weighed down by a feather weight placed on the opposite scale. Several similar but rougher representations of weighing the soul of a deceased person in the hall of Perfect Justice, and in the presence of Osyres, with an equal-armed balance, may be seen in the papyrus drawings on the wall of the stone staircase leading from the Egyptian sculpture gallery to the upper Egyptian rooms at the British Museum.

80. For the justness of an equal-armed balance it is requisite (1) that the two points of suspension of the pans from the beam be exactly in the same line as the centre of motion, or the fulcrum on which the beam turns when set in motion the line joining these three points is the axis of the beam; (2) that these two points be exactly equidistant from the centre of motion; (3) that there be as little friction as possible at the centre of motion and the points of suspension; (4) that the centre of gravity of the beam be placed a little below the centre of motion.

As to the theory of the relative positions of the centre of motion, the points of suspension, and the centre of gravity of an equal-armed balance, loaded with equal weights, it is to be remarked :—

(*a*) If the centre of gravity *coincides* with the centre of motion of the beam, and the three knife-edges are in the same right line, the beam of the balance will have no tendency to one position more than another, but will rest in any position in which it may be placed, whether the pans be suspended to it or not, and whether the pans be empty or equally loaded. Such a balance is wanting in proper action.

(*b*) If the centre of gravity of the beam when level be immediately *above* the fulcrum, the beam will upset with the smallest action ; that is to say, the end which is lowest will descend ; and it will descend with the greater velocity, according as the centre of gravity is higher, and the points of suspension less loaded.

(*c*) But if the centre of gravity of the beam be immediately *below* the fulcrum, the beam will not rest in any position but when level ; and if disturbed from that level position, it will vibrate, and at last come to rest in a horizontal position. Its vibrations will be quicker, and its tendency to the horizontal position stronger, the lower the centre of gravity, and the less the weight upon the points of suspension.

Again, as to the relative position of the central knife-edge on the fulcrum which determines the centre of motion of the beam, with the line joining the two outer knife-edges, which form the points of suspension, it is further to be remarked, (1), If the centre of motion be below the line joining the points of suspension, and these be loaded, the beam will upset unless prevented by the weight of the beam tending to produce a horizontal position, as shown in (*c*). In such case a small weight will produce an equipoise. In case of (*a*) a certain exact weight will make the beam rest in any position ; and all greater weights will cause the beam to upset, as in (*b*). (2), If the centre of motion be above the line joining the points of suspension, the beam will come to its horizontal position, unless prevented by its own weight, as in (*b*). (3), If the centre of gravity be nearly in the fulcrum, all the vibrations of the loaded beam will be made in lines nearly equal, unless the weights be very small, when they will be slower. The higher the fulcrum the quicker will be the vibrations of balances, and the stronger the horizontal tendency.

It is thus evident that the nearer the centre of gravity of the beam is to the centre of motion, the more delicate will be the balance, and the slower its vibrations. The tendency to a horizontal position is therefore increased by lowering the centre of gravity, in which case it will also require a greater additional weight to cause it to turn or incline to any given angle, and it is therefore less sensible with a greater load. The fixing of the centre of motion in a balance is consequently of peculiar importance, for on this depends the ease with which it will be affected by a smaller weight, and the readiness with which the beam will return to a horizontal position. And it will be seen that the best position of all is that in which the centre of motion is a little above the centre of gravity. Even in this, it should be proportioned to the distance of the weights from the fulcrum, and the amount of the load; and this object can only be attained in different beams by practice and experience. In order to regulate the centre of gravity in balances of precision, they are made to carry a small weight either over or under the centre of motion, which is movable by means of a screw.

From what has been said it would appear that if the arms of a balance be unequal, weights which form an equipoise will be unequal in the same proportion. But although for many purposes the equality of the arms of a balance is advantageous, yet a balance with unequal arms will weigh just as accurately as one with equal arms, provided the standard weight itself be first counterpoised, then taken out of the pan, and the weight to be compared be substituted and adjusted against the counterpoise. Or when proportional quantities only are required, they may be weighed against standard weights, taking care always to put these weights in the same pan. But in this case it is indispensable that the relative lengths of the two arms of the beam continue invariable. For this purpose either the three knife-edges should be truly parallel or the points of suspension and support be always in the same part of the knife-edge.

If the beam of an equal-armed balance be adjusted so as to have no tendency to any one position, as in (*a*), and the pans be equally loaded, then if a small weight be added to one of the pans, the balance will turn, and the point of suspension move with an accelerated motion, similar to that of falling bodies, but very nearly as much slower in proportion as the added weight is less than the whole weight borne by the fulcrum. The stronger the tendency to a horizontal position in a balance, or the quicker its vibrations (see (*c*) and (2)), the greater additional weight will be required to cause it to turn or oscillate to any given angle. If a balance were to turn with

$\frac{1}{10000}$th part of the weight, it would move at the quickest, 10,000 times slower than a falling body; that is to say, the pan containing the weight, instead of moving at the normal weight of a falling body, or through sixteen feet in a second of time, would fall only through $\frac{1}{50}$th of an inch, it would thus take about thirteen seconds to fall ¼ inch. Consequently all accurate weighing with a balance of precision, turning with a very small difference in the weights placed in the pans, must be slow.

81. In illustration of these principles of an equal-armed balance, the actual mode of construction of the several parts of a scientific balance of precision will now be described.

The fulcrum upon which the beam of a balance rests is a horizontal plane of polished steel or agate supported on the column of the balance. The beam itself rests upon this fulcrum, or bearing, by means of a knife-edge of hardened and polished steel. The line of the knife-edge is also in a horizontal plane at right angles to the axis of the beam. The whole length of the knife-edge ought consequently to be in contact with the plane of the fulcrum at every point. This knife-edge is at the lower edge of a steel prism, the section of which is an equilateral triangle; and the edge is ground with a small facet on each side, so as to increase the angle at the edge from 60° to 120°. Captain Kater, who was an eminent member of the Royal Society fifty years ago, and took the most active part, as a member of the Standards Commission, in directing and superintending the construction of the Imperial Standard

Weights and Measures legalised in 1825, considered the angle of 120° to be practically the best form for the knife-edge of a balance of precision, and the balances of the Standards Department have since been adjusted to this angle. It requires great care in a skilled workman to adjust the knife-edges, as the excellence of the action of balances depends very much upon them. The small facets on each side of the knife-edge ought to be exactly rectangular.

Whilst the central knife-edge is at the lowest part of the prism, the two knife-edges at the ends of the beam have on the contrary their edges uppermost, and it may be seen that the tendency of this arrangement is to bring the centre of gravity of the balance a little below the centre of motion. The two pans are suspended from these knife-edges by means of agate or steel planes bearing upon them. In order to preserve the nice adjustment of the knife-edges they are never allowed to rest upon their bearings, except when weighings are actually being made. At all other times the beam and pans are supported upon a frame attached to the column of the balance, but movable in a vertical direction upon it. When the balance is required to be put in action the support is very gradually lowered by means of a lever handle, and the knife-edges are brought upon their bearings so that the balance is free to act. As soon as the weighing is completed the supporting frame is again raised, and the knife-edges are thus lifted from their bearings.

The principal cause of discordances in the results of a successive series of weighings with a balance of precision, which should be guarded against, is the risk of the knife-edges not being brought again to exactly the same position on the plane bearings after the balance has been stopped and again set in action. The most perfect balance is that which varies least in the points of contact between the knife-edges and their bearings during successive weighings. For the attainment of this very important requirement the supporting frame is furnished at each of its extremities with two pins terminating in cones and made to fit exactly into corresponding conical holes in the plane bearings over the knife-edge at the ends of the beam. The two conical holes are in a line. with, and on each side of the knife-edge. The points of these four cones are all in the same horizontal plane. As the movement of the supporting frame in a well-constructed balance of precision is always in the same line, being guided by a vertical rod fitted to a cylindrically-drilled hole in the column of the balance, the knife-edges and their bearings should at all times be brought into contact in the same relative positions.

82. The pans of a balance of precision should be suspended in such a manner that in all positions the corresponding rods or chains of the two pans may be parallel to one another, else the weights, though equal, will not be in equilibrium. All the arrangements of the knife-edges and of the supporting frame may be seen in the following drawing of the kilogram balance of the Standards Department, constructed by Oertling.

To show the great importance of avoiding every risk of a displacement of the knife-edges on their bearings, by maintaining the most exact agreement between the position of the supporting frame and of

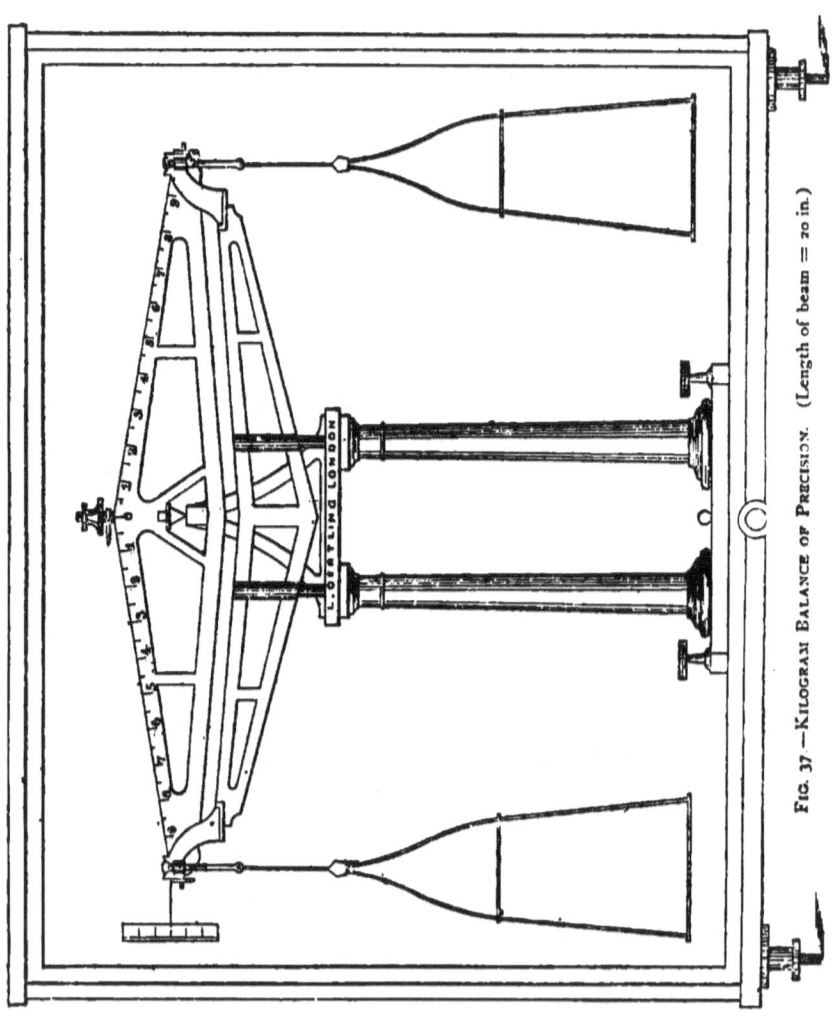

Fig. 37.—Kilogram Balance of Precision. (Length of beam = 20 in.)

the beam, it should be pointed out that in this balance the conical pins are a fixed part of the supporting frame, and must ascend and descend in the same vertical line, whilst the points of the conical holes, being movable with the beam, must move in the line of the circumference of a circle, the radius of which is the length of the actual line from the centre of motion of the balance to these points. There is consequently a risk of displacement from any such difference in the line of motion of the pins and of their conical holes, although the risk is diminished in proportion to the shortness of the beam. Such risk did not escape the observation of Captain Kater, and in some balances made under his direction by Mr. Bate, an arrangement was made by means of two side rods hinged to the central vertical rod, and of the proper radial length, to give the same circular motion to the supporting pins as to the conical holes.

83. There is a similar contrivance in the small balance recently constructed by Mr. Oertling for the Standards Department, from a design of M. Mendeleef of St. Petersburg, which deserves notice in other respects, and is represented in the following figure.

This balance, like that previously mentioned, is constructed to carry a kilogram in each pan, but the two balances vary materially in the length of their beams.

84. Long beams have been generally recommended because the quantity of motion in any part of a lever varies as its distance from the fulcrum; and therefore the greater distance of the points of suspension from the centre of motion in an equal-armed balance, the most distinguishable will be the motion arising from

For exacting accuracy the two weights must be on the same horizontal plane; to be at same distance from centre of Earth (for attraction) and for same density of atmos.

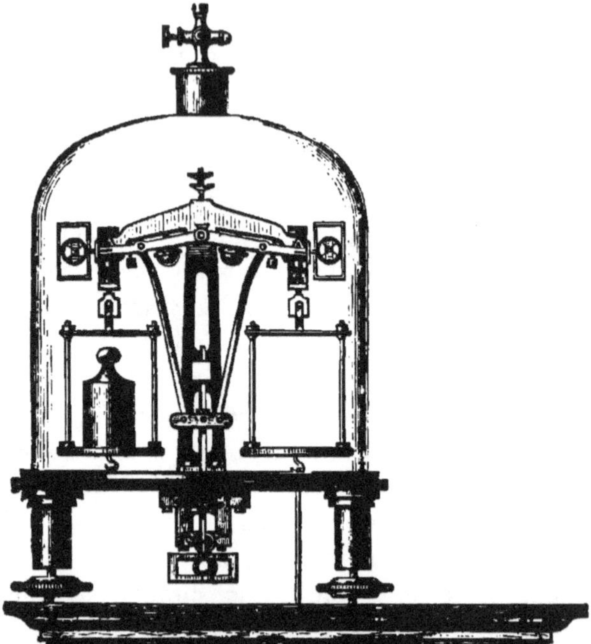

FIG. 38 — MENDELEEF BALANCE. (With short arms, and available as a Vacuum Balance.)

any small difference between the weights compared. On the other hand, there are certain advantages in the quicker angular motion, greater strength, and less weight of a short beam.

The larger balance has a beam of ordinary length, being twenty inches long, and is about two pounds in weight. Its sensibility is so great when in proper condition, that loaded with a kilogram in each pan, if a milligram, which is one-millionth part of a kilogram, and equal to $\frac{15}{1000}$ths of a grain, be added to one of the pans, it causes a difference in the resting point of the balance of about ten divisions of the index scale. One division, therefore, corresponds with 0·1 milligram.

The length of the beam of the Mendeleef balance is only $4\frac{3}{4}$ in., and its weight $\frac{1}{4}$ lb. This balance has been only very recently constructed, and its sensibility has not yet been sufficiently tested. But M. Mendeleef claims for his own balance, which served as a model for this, that one division of his scale is equivalent to 0·07 milligram.

An attempt is here made, and as it appears not unsuccessfully, to overcome the disadvantages hitherto considered to be inherent in a short-armed balance, of its oscillations being necessarily of very small extent, and minute differences in weights compared being consequently not observable. The mode adopted to meet these objections is to employ a microscope with a micrometer for observing the smallest movement of the pointer of the beam over a very fine graduated index. The result has been that with a kilogram in each pan, an additional weight of a milligram has

L

been found to cause an observed difference of fifteen divisions of the scale. This mode of observing the oscillation of a balance is, however, not new, as it was employed in the excellent balance constructed by Barrow, which was used by Professor Miller for all his weighings during the construction of the new Imperial Standard pound.

One great advantage of this short-armed balance is that being so small, and placed as it is upon a metallic plane, it can be covered with an ordinary glass bell receiver and used as a vacuum balance. In weighings in a vacuum it is very important to be able to exhaust the air rapidly and frequently, and that the balance case serving as a receiver should be as small as possible, so as to reduce to a minimum the volume of air to be exhausted.

85. It is to be observed that balances of precision ought to be enclosed in plate-glass cases, with a view not only to their preservation, but also to keep the balances as far as possible from being affected in their action by draughts of air, alterations of temperature, and other influences on their accurate action. In weighing with ordinary commercial balances the preponderance of either pan is indicated by a slender rod attached to the beam immediately over its centre of motion in a line perpendicular to the axis of the beam, and moveable freely between the two forks of the handle. It is called the tongue of the balance, and the degree of preponderance of either pan is shown by the greater or less deviation of the tongue from its normal vertical position. In balances of precision the preponderance is indicated by a longer

needle or pointer fixed to the beam, either at its centre, in a line perpendicular to the axis of the beam, and pointing downwards, or at either end, and in continuation of its axis. In both cases the pointer moves along a graduated scale. But an index-pointer placed perpendicular to the beam affects its equilibrium when turned from its horizontal position ; the measure of the momentum of the points being its weight multiplied with the distance of its centre of gravity from the vertical line. The error thence arising may, however, be corrected by continuing the index-pointer or counterpoising it, on the opposite side of the beam.

The finest balances of the Standards Department have the index-pointer at each end of the beam, as shown in Figure 37. For all weighings requiring special accuracy, the highest and lowest points reached by the needle in successive oscillations of the balance are read on the index scale through a telescope fixed at about five feet distance, by which means each reading can be satisfactorily taken by estimation to one-tenth of a division of the scale. In each case the mean between the highest and lowest reading is noted as the resting point of the balance. This will be more fully explained under the next head of the subject.

86. Amongst these balances of the Standards Department there is one balance of peculiar construction, mentioned above as having been used by Professor Miller for all his weighings during the construction of the new Imperial Standard pound, including his weighings of a kilogram. It has been lent to the International Metric Commission at Paris for the

weighings of the new standard kilograms. This balance was constructed by Barrow, and is similar in construction to Robinson's balances. The distance between the extreme knife-edges is 15·06 inches. The knife-edges work upon quartz planes. The middle knife-edge is about 1·93 inch long. Index scales marked upon thin and nearly transparent slips of ivory, a little more than half an inch long, are fixed to each end of the beam and oscillate with it. There are fifty divisions of the scale, 0·01 inch apart. The scale is viewed through a compound microscope, fixed in the glass case of the balance, having a single horizontal wire in the focus of the eye-piece. A glass screen is interposed between the observer and the front of the balance case. The mean value of the division of the scale was found to be about 0·002 grain, with one pound in each pan, and 0·005 grain, with a kilogram in each pan. All the results of his comparisons were noted by Professor Miller in hundredths of a division of the index scale of this balance.

87. In describing the mode of observing the oscillations of a balance of precision, attention should also be called to the method used by the late Dr. Steinheil of Munich, through the adoption of a principle originated by Gauss. The movements of the beam are observed by means of a small mirror placed immediately over the central knife-edge, and with its plane-surface normal to the axis of the beam. The observer is placed about twelve feet distant from the mirror and views through a telescope the reflection in the mirror of a vertical graduated scale placed close to the telescope. The ray of light reflected from the

mirror and thrown upon the graduated scale serves to indicate with great exactness any difference of the standard weights compared, by its angular deviation caused by the oscillation of the beam. The amount of this difference is ascertained by reading off the reflected scale the highest and lowest numbers at the turning points of the balance, as they coincide with a horizontal thread fixed in the telescope.

88. From the scientific construction of balances we next come to the scientific method of weighing. The ordinary method of commercial weighing by putting the weights in one scale and the commodity to be weighed in the other, and then observing when a sufficient equilibrium is produced, is inadmissible for scientific weighings, as it is subject to errors arising from defects in the balance itself. To avoid any such errors, and obtain scientific precision in the results, a check is required which is found in a system of double weighing, and in taking the mean result of a series of successive weighings. Two methods of double weighing are commonly used for the comparison of standard weights. One method, known as Borda's, and generally used in France, is that of *substitution*, or weighing separately two standard weights to be compared against a counterpoise placed in the other pan by placing them successively in the same pan. The difference between the mean resting points of the index-needle in these two weighings shows the difference of the two weights in divisions of the scale. The second method, known as Gauss's, but which was first invented by Le Père Amiot, and is now generally used in England and Germany, except for hydrostatic

weighings, is that of *alternation*, or first weighing the two standards against each other, and then repeating the weighings, after interchanging the weights in the pans. By this second method, no counterpoise weight is required, and *half* the difference between the mean resting points of the index needle shows the difference of the two weights, in divisions of the scale.

As the value of a division is continually liable to variation according to the condition of the balance, the state of the atmosphere, the weight in the pans, &c., it is necessary, for attaining very accurate results, to determine the value of a division for each comparison. This is done by an additional weighing after a very small balance weight, the value of which is exactly known, has been added to one of the pans, so that its effect on the reading of the index scale may be ascertained.

It is here to be observed that in all scientific weighings with balances of precision, the weights to be compared must be made so nearly equal that neither pan shall absolutely weigh down the other. The balance must merely oscillate so that the pointer does not move beyond the limits of the index scale. In order to obtain an equipoise within these limits, it is requisite to provide small balance weights, most accurately verified, to be added to either pan, as may be found necessary.

In cases where great accuracy is required, any number of successive comparisons may be made, with the object of taking the mean result of them all. This course not only lessens the probable error of the result, but is also a check against any accidental mistake either in noting the readings or in the computations. It is important to make the weighings in as short a tim· as possible, so as to avoid the risk of discordances in the results arising from variations in temperature, of moisture in the air, or other causes; and it is better to take only two or three comparisons at a time, and to repeat them on subsequent days, taking the mean result of all the comparisons.

89. The most accurate method generally adopted for noting the results of the oscillation of the balance by the movement of the pointer over the index scale, is as follows: it will also illustrate Gauss's method of weighing. The comparing standard, A, being in the left-hand pan, x, and the compared standard, B, in the right-hand pan, y, and sufficient equipoise being obtained by adding small balance weights, if requisite, the balance is put in action. The index is viewed through a telescope, placed at a few feet distance and adjusted to the index scale on the left-hand side of the balance, so as to observe the effect of the weight of A against B. The reading at the first turn of the pointer is disregarded. The three next turns are noted, and the balance then stopped. The reading at the third turn of the pointer, and half the sum of the readings at the second and fourth turns, are taken as the highest and lowest readings. Their mean is the resting point of the balance, or the reading of its position of equilibrium. The weights are then interchanged, when similar readings are taken of the weight of B against A, and dealt with in the same manner. These two observations constitute one comparison, and as before stated, half the

difference between the two resting points shows the difference of weight of A and B in divisions of the scale. An additional weighing is taken, after adding a small balance weight to either pan, in order to ascertain the value of a division; and if this balance-weight be added successively to each pan, the two weighings may be used as an additional comparison.

A similar mode of noting the weighings is pursued in a comparison by Borda's method, when for the two observations a counterpoise weight remains in the right-hand pan, y, and the two bodies A and B are weighed successively in x against this counterpoise, the difference between A and B being shown in divisions of the scale by the whole difference between the two resting points of the balance.

In using Gauss's method of weighing, it is very desirable to be able to transfer the pans and the weights contained in them from one end of the beam to the other without opening the balance case, and thus to avoid sudden changes of temperature of air within the balance case and consequent production of currents of air and discordances in the weighings. For this purpose, the following plan is adopted:—A grooved brass rod is fixed inside the balance case over and a little behind the beam. Upon this rod a brass slider is made to traverse by being attached to a slender brass rod drawn backwards or forwards from the outside of the case. A descending wire with a hook at the end is attached to the slider. For changing the weights, the slider and hook are brought to the right-hand end of the beam,

SCIENTIFIC USE OF INSTRUMENTS. 153

when the pan and weight are lifted from the beam and transferred to the hook by means of a brass rod curved at the end, and introduced through a small hole at the side of the balance case. The pan and weight are then slid to the other end of the beam, when the left-hand pan and weight are lifted in a similar manner from the beam, and the right-hand pan and weight substituted. It only remains then to transfer the left-hand pan and weight to the right-hand end of the beam.

This method possesses a further advantage. In making a great number of comparisons between two standard weights, they are exposed to some risk of being injured by wear, if they are taken up in the ordinary way with a pair of tongs. This risk is obviated by their being kept in the pans when lifted. Two light pans are used, of as nearly as possible equal weight, each of which has a loop of wire forming an arch with the ends attached to the opposite sides of the pan, so that it can be easily lifted with the curved end of a brass rod. The pans are marked x and y respectively. By interchanging the weights in the pans after a series of comparisons, and making a second series and taking the mean result, it gives the difference between the two weights, unaffected by any possible difference in the weight of the two pans. This contrivance is especially useful, when either of the weights to be compared consists of several separate weights. It was used by Professor Miller for all his more important weighings during the construction of the Imperial Standard pound.

The advantage possessed by Gauss's method of

alternation over Borda's method of substitution has been proved by Professor Miller as follows :—

Let A and B be two standard weights of the same denomination to be compared, and C the counterpoise of each.

For Borda's method, let the readings of the index be denoted by (A, C), when A is in the left pan and C in the right pan, and by (B, C), when B is in the left pan, and C in the right pan.

For Gauss's method, let (A, B) denote the readings when A is in the left pan and B in the right, and (B, A), when B is in the left pan and A in the right pan.

Let e be the probable difference between the recorded and the true position of equilibrium, that is to say, the probable error of a *single weighing* (not of a comparison, which requires two weighings).

Then by Borda's method, (A, C) has a probable error e, and (B, C) has a probable error e'; and the two weighings give the value of A − B with a probable error of $\sqrt{(e^2 + e'^2)} = e\sqrt{2}$.

By Gauss's method, (A, B) has a probable error e, and (B, A) has a probable error e; and the two weighings give the value of A − B with a probable error of $\frac{e}{2}\sqrt{2}$.

Thus the probable error of the result of two weighings by Borda's method is twice as great as by Gauss's method.

To obtain a value of A − B by Borda's method with a probable error of $\frac{e}{2}\sqrt{2}$, we must make four comparisons of two weighings each. Therefore one

SCIENTIFIC USE OF INSTRUMENTS. 155

comparison by the method of Gauss gives as good a result as four comparisons by Borda's method.

90. The following is an example of weighing by Gauss's method, and of the mode of recording the weighings :—

First Comparison of Platinum-iridium 1 *lb. P-i, with the Platinum Imperial Standard* 1 *lb. PS,* 21 *July,* 1874.

Compared	Compar- ing.	Readings of index scale in division.						Balance No. 3. Value of 1 divi- sion.	Com-⎫ ⎬=⎧ Com- pared⎭ ⎩paring.	
		Of compared.			Of comparing.				In divi- sions.	In grains.
		Highest.	Lowest.	Mean.	Highest.	Lowest.	Mean.			
P-i + y	PS + x	(1) 56	15	35·5	(2)30·3	62	46·15	gr. 0·002	− 5·2875	−0·01057
P-i + x	PS + y	(3) 63	10	36·5	(4) 50	44	47·00			
				36·0			46·575	Diff. = 2) 10·575		
									5·2875	
t. = 19°·62 C.		To find value of 1 div.		gr. 0·01 added to (3)				0·002		
b. = 755·09 mm. at 0° C.								0 0105750		
		(5) 50	13	31·5				(3) − (5) = 5)0·01		
									0·002	
∴ P-i = PS − 0·01057 gr. in air *t.* = 19°·62 C. *b.* = 755·09 mm. at 0° C.										

91. The result of these weighings gives only the apparent difference of the two standard pounds when weighed in air. In order to ascertain their true difference of weight, it becomes necessary to determine the weight of air displaced by each, from the data already mentioned (11), and to allow for any difference of weight of air displaced, according to the following formula :—

If the weights A and B appear to be equal in air, the weight of A − weight of air displaced by A is

WEIGHING AND MEASURING.

equal to the weight of B — weight of air displaced by B.

92. In determining the weight of ordinary atmospheric air so displaced in rooms where standard weights are compared, and containing a certain quantity of aqueous vapour and carbonic acid, the practice has been to take, as the unit of weight of air, a litre of dry atmospheric air free from carbonic acid, = 1·2932227 gramme, at 0° C., as determined by Ritter from the observations of M. Regnault in Paris, lat. 48° 50′ 14″, and 60 metres above the level of the sea, under the barometric pressure of 760 millimetres of mercury. Assuming that atmospheric air contains, on an average, carbonic acid equal to 0·0004 of its volume, the density of carbonic acid gas being 1·529 of that of atmospheric air, the weight of a litre of dry atmospheric air containing its average amount of carbonic acid, under the stated circumstances, is 1·2934963 gramme.

93. Allowance is next to be made for the difference of the force of gravity in latitudes other than Paris as well as for the difference of height of the place of observation above the mean level of the sea. Although the absolute weight varies with the latitude and with the height above or below the mean level of the sea, yet this variation is not felt in the comparison of standard weights in a vacuum, because the weights are equally affected on both sides of the beam. But in all weighings of standards in air requiring special accuracy, such variation must be taken into account in computing the weight of air displaced by each standard weight.

Mr. Baily has shown from his pendulum experiments[1] that if we take G to denote the force of gravity at the mean level of the sea in lat. 45°, the force of gravity in lat. λ, at the mean level of the sea

$$= G (1 - 0.0025659 \cos 2\lambda),$$

and Poisson[2] has proved that the force of gravity in a given latitude at a place on the surface of the earth at the height z above the mean level of the sea—

$$= \left\{ 1 - \left(2 - \frac{3\rho'}{2\rho}\right)\frac{z}{r.} \right\} \times \begin{array}{l}\text{(Force of gravity at the mean level of the}\\ \text{sea in the same lat.)}\end{array}$$

where r is the radius of the earth, ρ its mean density, and ρ' the density of that part of the earth which is above the mean level of the sea. If as is probable,—

$$\rho' : \rho = 5 : 11 : 2 - \frac{3\rho'}{2\rho} = 1.32 \text{ nearly}; \ r = 6366198 \text{ metres},$$

it follows that the weight in grammes of a litre of dry atmospheric air containing the average amount of carbonic acid, at 0°, and under the pressure of 760 millimetres of mercury at 0°, at the height z above the mean level of the sea in lat. λ is—

$$1.2930693 \left(1 - 1.32 \frac{z}{r}\right) (1 - 0.0025659 \cos 2\lambda).$$

At Cambridge, where Professor Miller's observations for determining the weight of the new standard pound were made, in lat. 52° 12′ 18″, about 8 metres above the mean level of the sea, and for which place his published Tables of Reductions were computed, the weight of a litre of dry air containing the average quantity of carbonic acid was found by him to be 1.293893 gramme. This weight of air is therefore a

---
[1] *Memoirs of the Astronomical Society*, vol. vii. p. 94.
[2] *Mémoires de l'Institut*, tome xxi. pp. 91, 238.

little greater than at Paris. From similar data, after taking a further correction by Lasch of the weight of a litre of dry air at Paris = 1·293204 gramme, the weight of a litre of dry air at Berlin (lat. 52° 30′, and 40 metres above mean sea-level) has been computed to be 1·29388 gramme.

The determination of the weight according to the latitude and height above sea-level, of a volume of dry air free from carbonic acid will be facilitated by the following Table issued by the Standard Commission at Berlin, and computed from the data here stated.

Correction for latitude and height above mean sea-level of the constant normal weight of a litre of dry air at 0° C. under a barometric pressure of 760 millimetres of mercury, = 1·29303 gramme, in lat. 45° at the mean sea-level; shown in units of the 5th decimal place:—

Lat.	Height in Metres above Mean Level.					
	0	50	100	150	200	250
40	− 58	− 59	− 60	− 62	− 63	− 64
41	46	48	49	50	52	53
42	35	36	37	39	40	41
43	23	25	26	27	29	30
44	− 12	13	14	16	17	18
45	0	− 1	− 3	− 4	− 5	− 7
46	+12	+10	+ 9	+ 8	+ 6	+ 5
47	23	22	21	19	18	17
48	35	33	32	31	29	28
49	46	45	43	42	41	39
50	58	56	55	54	52	51
51	69	68	66	65	64	62
52	80	79	77	76	75	74
53	91	90	89	87	86	85
54	102	101	100	98	97	96
55	+113	+112	+111	+109	+108	+107

The force of gravitation may thus be computed to be the same in lat. 46 at 450 metres above mean sea-level as it is in lat. 45 at sea-level. At the same time it is to be remarked that there is some uncertainty in the values here stated, arising from the fact of the actual flattening at the poles of the earth, and the distribution of its mass not being determined with mathematical accuracy, and from want of knowledge of the actual density of the crust of the earth, at the several places of observation.

The co-efficient of air under constant pressure between 0° and 50° C. is taken from Regnault's determination to be 0·003656 for 1° C.; in other words, between 0° and 50° C. the ratio of the density of air at 0° to its density at $t°$ is $1 + 0·003656\, t$.

94. With regard to the barometric pressure of the air and the allowance to be made for the pressure of vapour present in it, the density of the vapour of water is determined to be 0·622 of that of air; that is to say, the ratio of the density of the vapour of water to that of air is $1 - 0·378$.

Hence, if t be the temperature of the air, b the barometric pressure, v the pressure of the vapour present in the air, b and v being expressed in millimetres of mercury at 0° C., the weight of a litre of air at Cambridge becomes

$$\frac{1·293893}{1 + 0·003656\, t} \cdot \frac{b - 0·378\, v}{760}$$

The ratio of the density of air to the maximum density of water is found by dividing the above expression by 1,000, as a litre of water is the volume of 1,000 grammes of water at its maximum density. Professor Miller's Table I. gives the logarithms of this ratio at the normal barometric pressure of 760 millimetres, at the several degrees of temperature from $0°$ to $30°$. These logarithms require to be diminished only by 0·000026 for weighings at the Standards Office, Westminster, lat. $51° 30'$, and about 5 metres above the mean sea-level; and when diminished by 0·000132, they may be used for the reductions of weighings at Paris.

The values of the pressure of vapour at the same temperatures in millimetres of mercury at $0°$ according to Regnault's observations, are stated by Prof. Miller in a separate Table II. These values are given on the assumption that the pressure of vapour in rooms that are not heated artificially is two-thirds of the maximum pressure of vapour due to the temperature as shown by the results of experiments on the authority of Biot, Regnault, and Bianchi.

95. The actual mode of ascertaining the weight of air displaced by two standard weights may now be described, and illustrated by the reduction of the weighings in air of P-i and PS, already shown. For determining the temperature of the air and of the two standard weights during the weighings, two standard thermometers were placed in the balance case, and their readings noted at the beginning and end of the weighings. The density of P-i had previously been determined to be 21·42515, and of PS 21·15722. From these data, the weight of air displaced by each standard was ascertained by the following formula:—

Log. weight in grains of air displaced by P = log. h + log. At + log. $(1 + ePt)$ + log. weight of P in grains − log. Δ P.

Here t denotes the temperature of the air by the Centigrade thermometer;

SCIENTIFIC USE OF INSTRUMENTS.

b the barometric pressure of the air in millimetres of mercury at $0°$ C.;

v the maximum pressure of aqueous vapour contained in the air, also in millimetres of mercury;

$h = b - 0.378 \times \frac{2}{3} v$;

At the ratio of density of air at $t°$ to the maximum density of water;

ePt the allowance for expansion in volume of P, or the ratio of its density at $0°$ to its density at t;

ΔP the ratio of density of P at $0°$, to the maximum density of water.

By this formula, the required result was obtained as follows, the logarithms of the three first terms being taken from Prof. Miller's tables, pp. 785-791, of his account of the construction of the new standard pound, "Phil. Trans.," part iii. of 1856:—

Reduction of weighing in air $t = 19°·62$ C., $b = 755·09$ mm., in which P-i = PS - 0·01057 gr.:

	Compared Standard, P-i.	Comparing Standard, PS.
Log. h	2·875 5300	7·076 5295
,, At	4·200 9995	
,, 1 + ePt	0·000 2299	0·000 2299
,, weight in grains (7000)	3·845 0976	3·845 0980
	10·921 8570	10·921 8574
Deduct log. Δ	1·330 9238	1·325 4582
Result in log.	9·590 9332	9·596 3992
Whence, weight of air displaced, in grains	0·389 88	0·394 82
Excess displaced by PS	... ...	0·004 94

M

This excess of weight of air displaced by PS shows that P-i is comparatively heavier than PS in such air than it is in a vacuum, by 0·00494 gr. Therefore in a vacuum P-i = PS − 0·01551 grain.

96. The weight of ordinary air displaced by standard pounds and kilograms of various densities is shown in the following Table:—

	Metal.	Density.	Weight of air displaced.	
			$t = 62°F.$ $b = 30$ in. Imperial weight.	$t = 16\tfrac{2}{3}°\,C.$ $b = 761·986$ mm. Metric weight.
Imperial standard pound	Platinum	Δ 21·1572	grains. 0·403	—
	Brass	8·1430	1·047	—
	Bronze gilt	8·2829	1·029	—
Other avoirdupois pounds	Iron, adjusted with lead	7·1270	1·196	—
	Quartz	2·6505	3·217	—
	Glass	2·5179	3·385	—
French standard kilogram (= 2·2046 lb.)	Platinum	20·5487	—	milligrams. 59·25
English standard kilogram	,,	21·1379	—	57·60
French standard kilogram	Brass	8·2063	—	151·75
English standard kilogram	Bronze gilt	8·3291	—	146·23
Other kilograms	Iron, adjusted with lead	7·1270	—	170·84
	Quartz	2·6505	—	459·32

From these results, the proportionate weight of air displaced by a standard weight of any other denomination or density may be calculated without much difficulty; the weight of air displaced being in an inverse ratio to the density. Allowance for the difference of temperature or of barometric pressure may be computed from the following data :—

For a diminution of temperature of	10° F., deduct per cent.	2·12
,, ,,	10° C., ,, ,,	3·82
,, ,, barometric pressure of 1 inch	,, ,,	3·34
,, ,, ,, ,, 10 mm.	,, ,,	1·31

A corresponding addition per cent. to be made for increase of temperature or barometric pressure.

97. Reference has already been made (11) to the mode of ascertaining the volume or density of a standard weight by determining the difference of its weight in air and in water. The following practice for all such hydrostatic weighings was adopted by Professor Miller when determining the densities of all the standard weights constructed under the sanction of the Commission for restoring the Imperial Standards, and is also followed in the Standards Department. In this process it is requisite to employ pure distilled water; and with this object the water used in the Standards Department is twice distilled in a still of the best construction, erected in the office, and the best chemical tests are employed for ascertaining that the water is free from any foreign substances.

The vessel for containing the distilled water is a glass jar, rather more than 6 inches in internal height and diameter. A stout copper wire is stretched across the mouth of the jar (see Fig. 39) in such a manner as to leave a circular space in the middle, large enough to admit the passage of the standard weight P, the density of which is to be ascertained. This copper wire supports two thermometers, adjustable as to their height, for determining the temperature of the water at the mean height of P during the weighings. It also serves to sustain a glass tube, open at both ends, and placed close to the side of the jar. A small glass funnel is inserted in the upper part of the tube, and in the lower part are one or two pieces of clean sponge.

FIG. 39.—MODE OF HYDROSTATIC WEIGHING.

The standard weight P is suspended from a hook under the right pan of the balance, specially constructed for hydrostatic weighings. A fine copper wire, the weight of which per inch is known, is attached to the hook by a loop, and has another loop at the other end. To this lower loop is attached a stout wire, bent and terminating in a double hook, which fits round P, and holds it securely. The counterpoise of P is next placed in the left pan of the balance. The glass jar is placed under the right pan of the balance, P being suspended in it, and the water is gently poured into the funnel and the jar filled to the requisite height above P. The bubbles of air are arrested by the pieces of sponge, and, ascending up the glass tube are thus prevented from entering the jar. It is of importance to ascertain that no bubble of air is attached to P, and if so, it may generally be removed by the feather of a quill. But it sometimes happens that the weight P has an irregular surface, and air attaching to it cannot be thus dislodged. In such cases a small bell-shaped glass jar, just large enough to hold P and its supporting wire, is used. This vessel is filled with water sufficient to cover P, and is suspended over the flame of a spirit lamp by a stout wire, bent at its lower end into a ring, into which the jar descends to its rim, and the water is allowed to boil until it is seen that the air has been entirely expelled. When cooled, the small jar containing P is immersed in the water, which nearly fills the large jar and the small jar, with its wire, is then disengaged and lowered till P hangs clear of it, when it is removed. The transfer of P from the small to the large jar is thus effected without taking it out of the water.

For the actual weighing of P in water, after it has been counterpoised in air, weights equal to the difference of weight of P in water and in air are placed in the right pan till equilibrium is produced, when the readings of the scale are observed. P is next removed, leaving its hook suspended in the water, and a volume of water equal to the volume of P is added to the water in the jar, so as to leave the same quantity of wire immersed as before. The requisite weights are then added to the right pan, until the equilibrium, which has been disturbed by the removal of P, is again produced, when the reading of the scale is observed and noted. This gives the actual weight of P in water.

The thermometers in the water are so placed as to give the temperature of the water at the centre of gravity of P. Another thermometer is placed in the balance case to give the temperature of the air during the weighings. The reading of the barometer is also noted.

Having determined the weight of P in air of ascertained density, its volume and density are calculated according to the following formula, the unit of volume being the volume of a grain weight of water at its maximum density:—

Let P in water at $t°$ appear to weigh as much as Q in air. Then the weight of water at $t°$ displaced by P = weight of P − weight of Q + weight of air displaced by Q.

Log. volume of P = log. weight in grains of the water displaced by P + log. W_t − log. $(1 + eP_t)$; where W_t is the ratio of the maximum density of water to its density at t, and ePt is the expansion in volume of P at t. (The logarithms of these values are given in tables.)

Log. density of P = log. weight of P in grains − log. volume of P.

SCIENTIFIC USE OF INSTRUMENTS. 167

As the true weight of P, which is its weight in a vacuum, cannot be ascertained until its volume or density is known, an approximate value of the volume and density of P may be found by assuming the weight of P to be equal to its apparent weight in air; and this value of the volume of P may be used in reducing the weight of P, and thus a more accurate value of the volume and density and true weight of P may be found. This process should be repeated when greater exactness is required.

98. The actual process will be more easily understood by the following example of determining the volume and density of a bronze lb., No. 35, in the Standards Department (see Appendix to Fifth Report of Standards Commission, p. 68):—

		Grains.
From weight of 1 lb., No. 35, in air, $t = 64°·85$		
F. $b. = 29·915$ in. $=$	... ...	7000·00000
Deduct weight of 1 lb., No. 35, in water, $t =$		
60°·09 F. $b_1 = 29·915$ in. $=$	... ...	6143·91825
		856·08175
Add weight of air displaced by 6143·91825 gr. of bronze $=$	... ...	0·87335
∴ weight of water displaced by 1 lb., No. 35 $=$		856·95510

Log. weight of water displaced $=$	2·9329580
Add log. Wt. $=$	0·0004108
	2·9333688
Deduct log. $(1 + eP t) =$ ...	0·0003465
∴ log. vol. 1 lb., No. 35 $=$ ...	2·9330223
Deduct from log. weight in air of 1 lb., No. 35 (in grains) $=$ ...	3·8450980
∴ log. Δ lb., No. 35 $=$	0·9120757

Whence vol 1 lb., No. 35 $= 857·008$ grains
weight of water; and Δ 1 lb., No. 35 $= 8·16724$

This result is obtained by taking the nominal weight of 1 lb. No. 35 in air = 7000 gr. as its true weight. A slight correction might be made, if very great accuracy were required, by a more precise computation of its true weight, taking its density = 8·16724, and by repeating this process, substituting its corrected density.

The density of a specified weight may be found approximately by dividing its weight in air by the weight of water displaced by it; thus

$$\frac{7000}{856\cdot9551} = 8\cdot16834.$$

99. In the comparison of standard weights, the difficulty and risk of error in determining the weight of air displaced by them is to be avoided by weighing them not in air but in a vacuum. Two methods are employed for this kind of weighing.

In the first and simplest method, when an ordinary balance of precision is used, each standard weight is placed in an exhausted receiver just large enough to hold it, and is weighed separately against a counterpoise by Borda's method. Sensible discordances have, however, been found in the results of this method of weighing in exhausted receivers, which render its use inexpedient when scientific accuracy is required. These discordances are perhaps attributable to a small quantity of air being present in the receiver during the weighings, the amount of which cannot be accurately determined. Another probable cause is a change in the temperature and atmospheric pressure affecting the balance itself and the weights in the pans during the long time necessarily occupied in the whole pro-

SCIENTIFIC USE OF INSTRUMENTS. 169

cess of this method of weighing. It may be generally stated as a rule that the risk of discordances in the results of weighings is in proportion to the time occupied in the operation. Such discordances are not found when the weighings are made by the second method, when a vacuum balance is used, that is to say, when the balance case itself is made an exhausted receiver.

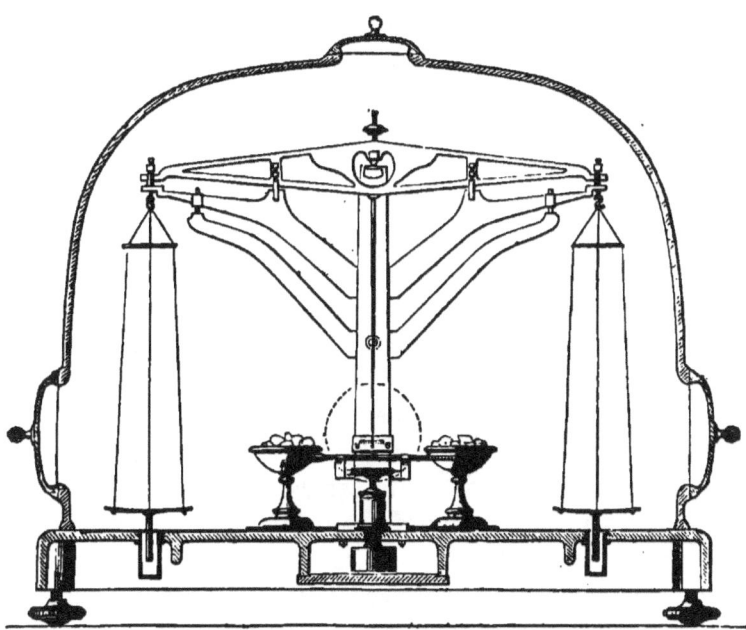

FIG. 40.—DELEUIL'S VACUUM BALANCE.

100. A vacuum balance has been constructed at Paris by M. Deleuil, and is now used at the Conservatoire des Arts et Métiers, consisting of a balance of the best construction placed in a very strong cast-iron

case that can be made perfectly air-tight. This case has four circular openings for giving admittance and light to the inside, which are closed with strong glass covers. It has a stuffing box for the handle of the lever by which the balance is put in action and arrested. This balance has been found to give very accurate results of weighing in a vacuum. But the comparison of standard weights in this vacuum balance takes a considerable time, from the necessity of opening the case and re-establishing a vacuum, at least a second time, even when Borda's method is used, in order to change the weights in the pans. This must be done again if a small additional balance weight is required to be placed in either pan for obtaining a sufficiently approximate equilibrium, so that the pointer may not exceed the limits of the index scale. It will be seen that the needle point is placed in a vertical position, moving along an index scale at the lower part of the column, according to the usual construction of balances of precision in France.

101. Some improvements on Deleuil's vacuum balance have been designed by Professor W. H. Miller, and have been practically carried out in a vacuum balance constructed by Mr. Oertling for the Standards Department. The balance case consists of a strong brass frame cast in one piece, with a rectangular base, two sides, and an arched top. Two solid glass plates, each $1\frac{3}{4}$ in. thick, form the front and back of the case, being clamped to plane surfaces of the brass frame, and made air-tight by interposing thin india-rubber. They are thus removable when required, for instance, when any alteration is needed in the balance. There

is a circular opening $4\frac{5}{8}$ in. in diameter, on each side of the brass frame, similar to those on Deleuil's balance, to which glass covers are fitted. No stuffing box is employed, but when the standard weights to be compared are placed in the pans, and the balance case exhausted, contrivances are arranged for putting the balance in action and arresting it, for adding any balance weights to either pan and removing them, and for interchanging the pans and weights by transferring them to the other end of the beam, without any disturbance of the vacuum, or necessity of opening the case.

These arrangements enable the weighings to be made by Gauss's method of alternation. The balance case is firmly placed upon a strong mahogany stand. Two iron tubes are fixed underneath and opening into the balance case. They rest in iron cups containing a sufficient quantity of mercury. Within each tube is a steel rod rising to the required height inside the balance case, and having at the top an arm of convenient form. By means of a simple lever handle outside the tube, either rod can be lifted about an inch, and it can also be turned round. By this rotary motion, when the left-hand rod is in its normal position, it acts upon two bevelled wheels, and thus lowers the supporting frame of the beam and puts the balance in action ; and by reversing the motion the action is stopped. By raising either rod to nearly its full height, it can be made to take up one of several small balance weights riding on a little rail fixed to the pillar of the balance, and transfer it to a similar rail at the top of the pan, or to transfer it back again.

Again, by raising either steel rod to an intermediate height, and turning its arm under the arched rods of one of the pans, and then raising it a little, the pan and weight can be lifted off the hook of the beam and transferred to a small carriage standing upon a railroad near and parallel to the front of the balance case. In a similar way the other pan and weight can be transferred to a second carriage at the back of the case. By means of a cord and pulleys, one of which is upon the right-hand steel rod and can thus be turned round with the hand, the two carriages can be moved to the other ends of the case, and then each pan with its weight can be attached to the hook at the other end of the beam. The desired results are all thus attained, and the whole action of the balance is open to view.

The construction of this new vacuum balance may be seen from the annexed figure.

The balance itself is similar in construction to the other standard balances made by Mr. Oertling. It is constructed to weigh a kilogram in each pan. There are two standard thermometers inside the case, one fixed to each pillar, and adjustable as to height, so that its bulb may be on the same level as the centre of gravity of the weight. A mercurial gauge is fixed between the pillars, and there is the same arrangement of three tubes and stop-cocks communicating with air-pumps and with a manobarometer, as in Deleuil's vacuum balance. Two glass vessels containing chloride of calcium, are also introduced for absorbing any moisture in the balance case.

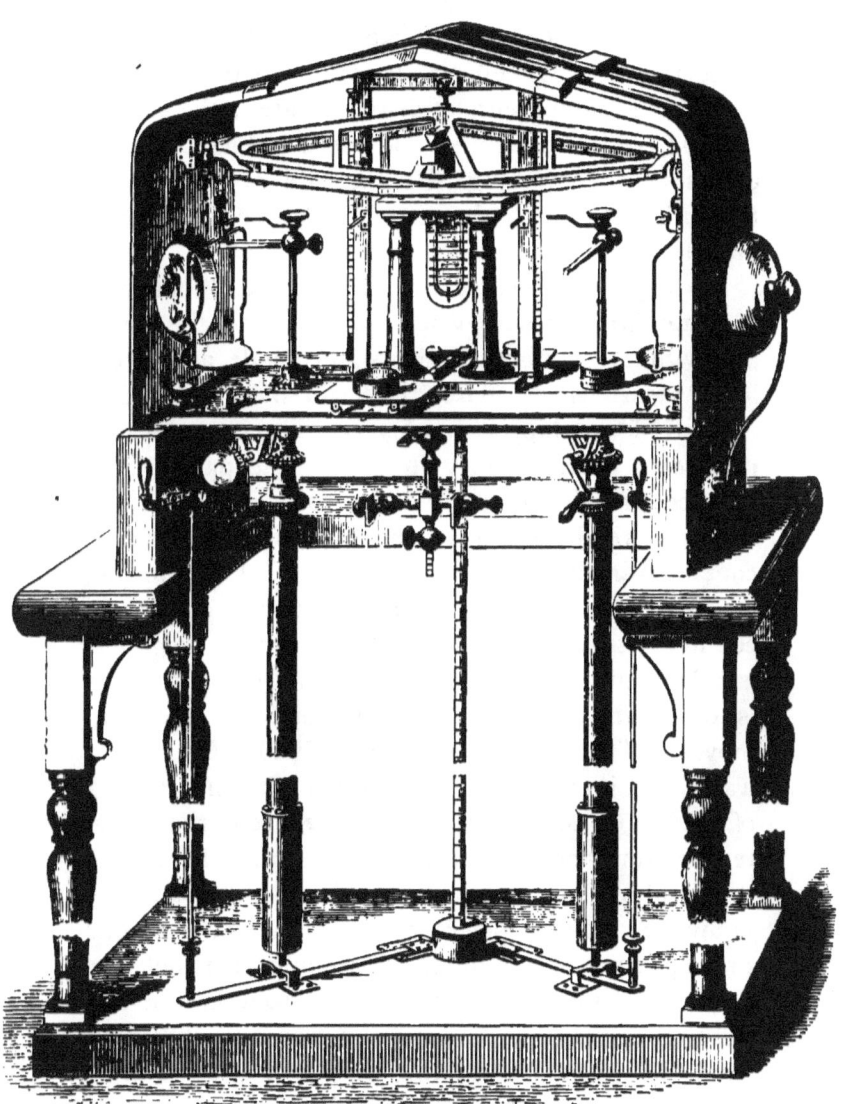

Fig. 41.—Vacuum Balance of Standards Department (⅛ size).

102. There are two ways of comparing and verifying standard measures of capacity. The first and most accurate, as well as scientific method, is by weighing their contents of distilled water ; the second method, which is simpler and more ordinarily used, consists in comparing the measure of water contained in them, with the contents of a verified standard measure.

In weighing the contents of distilled water contained in a standard measure, when quite full to the brim, and with the surface of the water made accurately level by a disc of plate glass slid over it, Borda's method of weighing is employed. The measure with its disc is placed empty in one of the pans of the balance, and is accurately counterpoised. A verified standard weight equal to the legal weight of water contained in the measure is then added to the pan containing the measure and disc, and is accurately counterpoised, and a sufficient number of weighings is taken until the mean resting-point of the balance is determined. The standard weight is then removed. The measure is exactly filled with distilled water, and its temperature, together with the reading of the barometer, noted. Any difference in the actual temperature and barometer pressure from the normal temperature and pressure is to be compensated by equivalent weights placed either in the measure pan or weight pan as required. If an equipoise is not now obtained, additional weights are placed in the pan until an equilibrium is produced, and any difference from the normal correcting weight for temperature and barometric pressure, either plus or minus, shows the error of the measure in relation to its legal weight of water at the standard temperature and barometric pressure.

103. For ascertaining the exact amount of the proper corrections for temperature and barometric pressure, authoritative tables are computed both for imperial and for metric measures. Such tables will be found in the papers appended to the Fifth Report of the Standards Commission, published in 1871 (pp. 81, 193, and 196), and to the Sixth Annual Report of the Warden of the Standards, published in 1872 (pp. 49 and 51).

The following abstracts of these tables may serve by computation for other temperatures and barometric pressures.

Corrections for temperatures differing from 62° Fahr. in the imperial gallon of water, when weighed in a brass or bronze, or a glass measure, against brass or bronze weights; and also for barometric pressure differing from a mercurial column of 30 inches :—

	For temperature.		For barometric pressure.	
t.	In brass or bronze measure.	In glass measure.	b.	
°F.	Grains.	Grains.	In.	Grains.
50	+ 35·50	+ 47·21	29	+ 2·51
55	+ 25·34	+ 32·17	29·3	+ 1·76
60	+ 8·50	+ 10·45	29·6	+ 1·00
65	− 14·57	− 17·51	29·9	+ 0·25
70	− 43·47	− 51·28	30·2	− 0·50
75	− 77·75	− 90·43	30·5	− 1·26

+ signifies that the stated weight is to be added to the brass or bronze weights.
− that the stated weight is to be added to the pan containing the measure of water.

Corrections for temperatures for the weight to be added to a metric litre of water in either a brass or a glass measure, when weighed against brass weights, the barometer being at 760 millimetres:—

$t.$	In a brass measure.	In a glass measure.
	Gramme.	Gramme.
8° C.	0·805	1·005
12	0·925	1·224
15	1·058	1·523
16·6 (=62° F.)	1·320	1·738
18	1·479	1·929
21	1·906	2·430

These corrections for a litre are to be increased or diminished 1·4 milligram for every 1 mm. of barometric pressure differing from 760 mm. Both tables of corrections are based on the difference of the weights of the specified measure of air and of water, according to the variation of density from temperature, barometric pressure, &c., and on the difference of volume of water in a brass or glass measure, according to the variation of capacity from the temperature.

In the determination of standard measures of capacity by the weight of their contents of water, it is evident that whilst the capacity of the measure is increased by its expansion when at a higher temperature than the normal temperature (for instance, than the litre at 0° C.), so the weight of the bulk of water contained in the standard measure when at the normal temperature of the maximum density of water (or 4° C.), is diminished when the water is of a less density at a higher or even somewhat lower temperature. But there will be one temperature of which these two elements of difference compensate each other, and this compensating temperature will vary according to the rate of expansion of the measure. According to Van der Toorn, as quoted by Mr. Hassler, in his papers on the Standards of the United States of America, the temperature at which by the difference of expansion of the water and of the litre measure, the litre again contains an equal weight of water as at the maximum density, is as follows:—

For a glass vessel	45° Fahr or	7°·22 C.
,, brass ,,	52·8 ,,	11·56
,, copper ,,	51·8 ,,	11·00
,, pewter ,, (5 tin 1 lead)...	56·3 ,,	13·50

No mention is made of the rates of expansion upon which these results are based. Taking the rates of expansion now adopted in the Standards Department the compensation temperatures are as follows:—

For a glass vessel	50·0 Fahr. or	10°·0 C.
,, brass ,,	57·4 ,,	14°·1

104. The other method of comparing measures of capacity, and the one most commonly used, is by comparing the measure of water contained in them, when quite full to the brim and accurately levelled by means of a disc of plate glass, with the measure of water of equal temperature similarly contained in a standard measure of the same metal. Sufficiently accurate results for all ordinary purposes may be obtained by this last mentioned process, if care be taken to leave

the two measures to be compared in the same room with the water for at least twenty-four hours beforehand so that they may attain the same temperature; to have the upper rim of each measure in a true horizontal level; to remove any air-bubbles from the measure of water; and when transferring the water from one measure to the other, by means of a syphon, to see that both the syphon and the second measure are previously wetted, so that no error arise in the results of the comparison from any difference in the small quantity of water adhering to the sides of either vessel.

105. With regard to comparing instruments for standard measures of length, their construction has necessarily varied according to the form of the standard measures to be compared. As already stated, the earlier scientific standards of length were defined by two points, and all comparisons were made by means of a beam compass.

Prior to the introduction of the use of the micrometer microscope by Mr. Troughton, the beam compass with points was the only mode by which the distance between two given lines or dots could be ascertained and set off on another scale, or by which their differences could be determined. The beam compass consisted of a wooden beam, with two steel pins attached to it at right angles, one of them fixed, the other movable. The points of the pins were set upon the given divisions of a scale, and were then transferred to the corresponding divisions of another scale. Their difference, if any, was ascertained or estimated by a diagonal scale. In fact there were no means of ascertaining any minute differences, and the parties satisfied themselves with a merely approximate comparison. And necessarily so after many experiments, since the points of the compass would wear away the edges of the dots, by repeated application to the same, or nearly the same, spots. We find this to have been the case in some of the most important standard scales of former times.

This disastrous result led to the discontinuance in this country of defining standard measures of length by points or dots, and to the substitution of fine lines. Such line-standards, or measures à traits, as they are tèrmed, have hitherto been little used abroad, where end-standards, or measures à bouts, have been generally adopted.

106. The introduction of the use of the micrometer microscope was a great step in advance towards the attainment of scientific accuracy in the comparison of our standard measures of length. It enabled optical observations to be made without injurious contact to the defining points or lines, and thus without interference with the permanence of the measures. Several descriptions of comparing apparatus with micrometer microscopes have been constructed at various times, but all are made upon the same principle. The microscope is fixed in a vertical position, and is provided with a spirit level and with screws for accurate levelling and focal adjustment. The defining marks of the two standard measures to be compared are brought successively

under it, their height being adjusted to the focal distance of the microscope. Any difference of length between the defining marks of the two measures is read off from the graduated head of the micrometer. This part of the apparatus consists of an endless screw with the very finest threads, having a large head divided into 100 parts. The screw is placed in a horizontal position, and when turned carries with it a nut moving in horizontal guides, and furnished with cobweb lines stretched across an open frame. Two of these lines (*b b* Fig. 42) cross each other at equal angles to the axis of the screw, and so that a line bisecting them is normal to its axis. Two other lines (*c c*) are placed nearly close, and parallel to each other and normal to the axis of the screw; and there are two longitudinal lines (*d d*) parallel to the axis of the screw, by means of which this axis is made parallel to the axis of the measure under observation. When turning the screw, the number of revolutions is read off from a rack (*a*) at the edge of the open frame and parallel to the screw, and the number of divisions in one revolution from the micrometer head by the aid of a fixed line marked on the upper surface of the microscope. Looking through the microscope at the magnified first ten hundredths of the inch 36—37 marked on the subdivided standard yard of the Standards Department (here inverted), the field of the microscope is seen as represented in the annexed figure.

In this figure the cross lines are used for observation, and are seen adjusted to the defining line of 0·03 in. The indicating mark at the rack shows the screw to be turned between one and two revolutions from the middle of the field.

107. All micrometer microscopes used for the comparison of standard measures of length are constructed upon the principle thus described. But there are various kinds of arrangements for supporting the

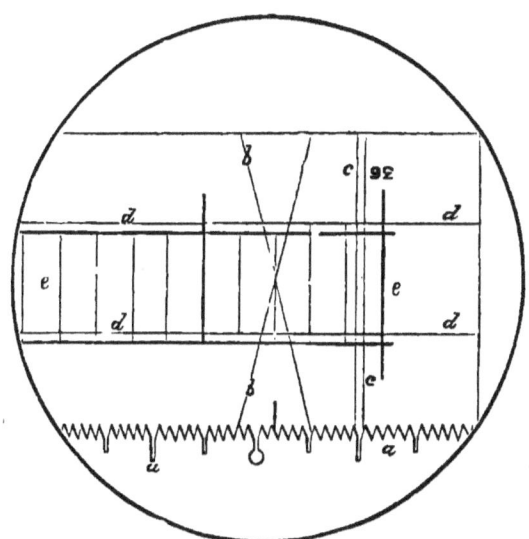

FIG. 42.—FIELD OF MICROMETER MICROSCOPE.

standard measures in a proper position, and for more conveniently bringing their defining marks under the microscopes. Under one of the arrangements, a single micrometer microscope is used, being fixed over the supporting apparatus, which, for the purpose of comparison, has both a transversal and a longitudinal displacement.

The two standard measures (denoted as A and B) being placed with their axes exactly parallel, and their defining marks as nearly as possible in the same line normal to the axes, the left-hand defining mark of A is brought under the microscope, and the position of the micrometer read off on the index scale and noted. By the transversal displacement, the left-hand defining mark of B is next brought under the microscope, and the reading of the index scale noted. The same result may be produced without having recourse to a transversal displacement by using a double microscope diverging from a single eye-piece. The two measuring bars are then moved their whole length by longitudinal displacement, and the right-hand defining marks of A and B successively are read off and noted, thus affording the means of ascertaining the difference of length of the two standard measures. The temperature of the bars at the beginning and end of the observations must be determined by thermometers, and the mean temperature noted; and allowance must be made by computation for any difference of length arising from unequal expansion or contraction of the two bars, when this temperature differs from the standard temperature. For this purpose it is absolutely necessary that the coefficient of expansion of each standard bar be previously determined.

The following Table shows the absolute expansion of a yard bar of various materials, for 1° F. and for 30° F., and of a metre bar for 1° C. and for 30° C., respectively, at the rates of expansion now generally adopted :—

EXPANSION OF YARD AND METRE BARS.

	Yard.		Metre.	
	For 1° F.	For 30° F.	For 1° C.	For 30° C.
	in.	in.	mm.	mm.
Platinum	0·000171	0·00514	0·00476	0·14280
Brass	0·000344	0·01032	0·00956	0·28698
Bronze	341	1023	947	0·28410
Copper	314	942	873	0·26180
Wrought Iron ...	220	660	550	0·16500
Cast Iron	198	594	611	0·18333
Cast Steel	207	629	575	0·17250
Glass	177	532	492	0·14766
Pine Wood ...	0·000099	0·00297	0·00275	0·08265

108. The method of comparing with a single microscope is used in France, but not in England, where the risk of error arising from the longitudinal movement of the bars is avoided by using two microscopes, and only a transversal displacement of the bars during the observations, although there are means of longitudinal displacement for the purposes of adjustment. The objection raised against the use of two microscopes, that the distance between them may vary during the period of observation by the expansion or contraction from alteration of temperature of the material which unites them, is obviated by fixing them firmly and independently upon a solid stone support.

109. The micrometrical apparatus specially made for the comparison of the Imperial standard yards constructed by the Committee for Restoration of the Standards, consisted of two microscopes carrying micrometers, and firmly supported upon large squared stones built upon a massive foundation of masonry.

This apparatus was used by Mr. Sheepshanks for all the comparisons of the standards yards, the microscopes being fixed as nearly as possible at the distance of a yard apart.

The bars when under comparison were placed in an iron trough or drop-box with double sides and bottom, and were surrounded by a large mass of water for the regulation of the temperature. The drop-box contained quicksilver upon which the bars floated, and was supported upon a carriage or travelling platform by which the bars under comparison were successively placed under the microscopes, and brought by adjusting screws to the right focal height. Two lamps for throwing light on the prisms of the microscopes and thus illuminating the defining lines of the standards under comparison, were placed so that the line passed through perforated holes in the stone supports of the microscopes. The bars intended to be floated on quicksilver were all covered with gold-beaters' skin.

This micrometrical comparing apparatus was designed more immediately for the line standard yards, by direct observation of their defining lines through the microscopes, and an opinion is expressed in the Report of the Commission that no apparatus possessing equal firmness and delicacy, or ensuring equal immunity from irregular thermal expansion, had ever before been used for the comparison of standards *à traits*. By means of this apparatus and by originating a new system of the most accurate thermometers, the Commission were enabled to overcome all difficulties arising from continued altera-

SCIENTIFIC USE OF INSTRUMENTS.

tions in the length of metallic bars caused by changes of temperature, and to determine the lengths of all the standard yards verified by them within the variation caused by a change of temperature of $0°\cdot01$ F.

110. The apparatus was also made available for the comparison of end standard yards with line standards by means of the following process, which originated in a suggestion from the Astronomer Royal. For such comparisons, a well was sunk about the middle of each end-bar to its mid-depth, and marked with a transverse line upon a gold pin, as in the line measures. By placing two such end-bars in a line end to end, it is obvious that the two inner half yards will make together a line-standard yard, comparable under the microscopes with an ordinary line-standard yard. Then by interchanging the positions of the end-bars so as to bring their other ends in contact, the length of the other two halves may in like measure be compared with a line standard. Adding together the two measured distances, we have a determination of the sum of the two end-bars by doubling the length of the line-bar.

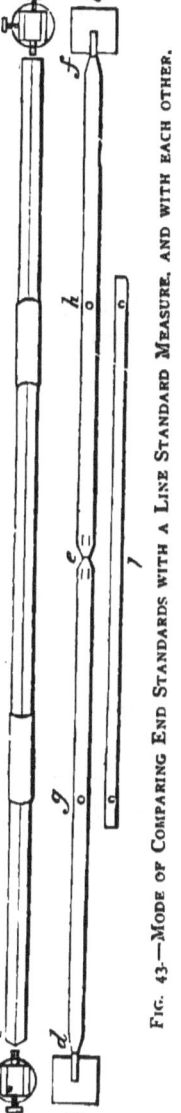

FIG. 43.—MODE OF COMPARING END STANDARDS WITH A LINE STANDARD MEASURE, AND WITH EACH OTHER.

In order to complete the determination, it was necessary further to ascertain the difference between the whole lengths of the two end-bars. This was effected by applying the above described process to *three* end-bars, and determining the measure of the sum of each pair of end-bars in relation to double the length of the line-bar. Three equations are thus given to find three unknown quantities, and the length of each of the three end-bars is thus found without difficulty.

Fig. 43 (on preceding page) shows the arrangement of two end-standard yards for comparison with a line-standard yard, according to the Astronomer Royal's method :—

Here $b\ b'$ are two rectangular contact pieces at the outer terminations of the end-standard yard $d\ e$ and $e f$ pressed together by springs. Each of these end-standard yards has defining lines $g\ h$ marked at the middle and mid-depth of the b r. The line-standard yard l is placed so that its length can be readily compared with $g\ h$ under the microscopes.

Fig. 43 illustrates, also, the mode of comparing one end-measure with another, or with two end-measures in a continuous line. Each contact piece has a fine transverse line marked on its upper surface, very near to the point of contact, the actual distance being accurately determined, and these lines are compared under the microscopes.

The comparing apparatus, with its two micrometer microscopes, used by the late Mr. Sheepshanks, has been removed from the lower cellar under the apartments of the Astronomical Society at Somerset House, where it was used by him, to the strong room of the Standards Office, and is put into complete order for use. It is thus preserved as an historical scientific instrument in connection with the construction of the Imperial standard yard.

111. Another and a more complete apparatus on the same principle has been supplied to the Standards Department by Messrs. Troughton and Simms, and is used for the departmental comparisons. It is placed in a large room on the first floor of the old Norman Tower, a position that possesses the advantages of great stability and little variation in temperature. A strong stone shelf is fixed to the solid wall and extends for a length of nearly 24 feet, upon any part of which the microscopes can be placed. The supporting frame of the measuring bars is a metallic plane resting upon a carriage somewhat similar to that in the Sheepshanks apparatus, but which moves longitudinally in a line exactly parallel with the front of the stone shelf on wheels running upon three iron rails firmly bolted to the levelled stone floor. This arrangement enables comparisons to be made of any standard bars up to 20 feet in length.

112. In comparing two measuring bars it is objectionable to place them directly upon a plane support. It has been proved that there is a risk of discordances in comparisons being caused by almost undiscoverable inequalities in planed surfaces, as well as by a difference of temperature in the plane surface and the under surface of the measuring bar, when thus placed. To guard against this risk, the bars are supported upon rollers, and the measuring bars ought to be stiff enough

188   WEIGHING AND MEASURING.

to bear to be supported upon a few points at which rollers can be conveniently applied. For a short bar two rollers are sufficient; for a longer bar more supports are required. The standard yard bars are supported upon eight rollers, and it is always requisite that each support should exert the same vertical pressure upwards, in order that the interval between two points upon the surface of the bar may not be altered by the flexure. It is easily seen that an arrangement of levers by which equal pressure upward may be exerted at four or eight points is very simple. Each bar rests upon two brass lever-frames. The form of lever-frame, with four rollers, is shown in the following figure:—

Section

FIG. 44.—COMPOUND LEVER FRAME FOR SUPPORT OF STANDARD BARS.

A metallic base plane a supports the centre of motion b at the centre of a lever of first order. This lever consists of two parallel plates c c, separated by a space much larger than the breadth of the measuring bar, and joined by cross-ties. The lever carries the two centres of motion, $d\, d$, of two similar but smaller levers of second order, each consisting of two parallel plates, falling between the parallel plates of the lever of first order, but also more widely separated than the breadth of the measuring bar. Each lever of second order carries two rollers upon which the standard-bar *s* rests. It will be seen that the pressures on the four different rollers when carried by one lever-frame are necessarily equal; and if two similar lever-frames are

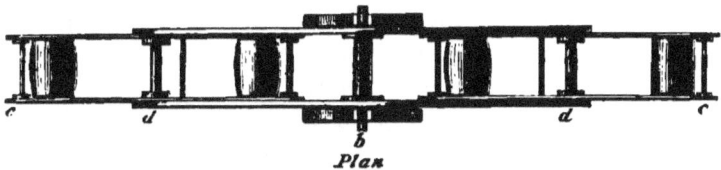

FIG. 45.—LEVER FRAME, WITH ROLLERS FOR SUPPORTING STANDARD YARD.
(⅓ size.)

placed symmetrically under the bar, the pressures on the eight different rollers are necessarily equal.

It has been shown by the Astronomer Royal, in his paper printed in the Royal Astronomical Society's Memoirs, vol. xv., that the value of the intervals (supposed equal) which ought to exist between different supports of a bar, each support exerting the same vertical pressure upwards, so that the interval between two points upon the surface of the bar may not be altered by the flexure, is as follows: n being the number of supports, the resulting intervals of supports is :—

$$\frac{\text{length of bar}}{\sqrt{(n^2 - 1)}}.$$

Thus the length of the standard yard bars being 38 in., and eight supports being required, the resulting interval of support in inches is

$$\frac{38}{\sqrt{63}} \text{ or } \frac{38}{7\cdot 937} = 4\cdot 78 \text{ in.}$$

113. In order to ascertain with scientific precision how far the results of comparisons of standards obtained by the use of weighing and measuring instruments are to be depended upon for their accuracy, a calculation is to be made of the probable error of every such result, whether it be the result of a single comparison, or the mean result of any number of comparisons. And when other elements are to be taken into account, it is necessary that the probable error of each computation should be determined and allowed for before the final results of comparison can be determined and allowed for.

The mode generally adopted for calculating the probable error is based upon the method of least squares, and is fully stated by the Astronomer Royal, in his "Theory of Errors of Observation," pp. 44-7.

To determine the Probable Error.

I. Find the mean of all the results of the observations.

II. Take the excess or deficiency of each result from the mean of the whole as found in I., regardless of signs, which will show the apparent error of each result.

III. The probable error of the result of any single observation

$$= 0.6745 \times \sqrt{\frac{\text{sum of squares of apparent errors}}{n - 1.}}$$

IV. The probable error of the mean results of all the observations

$$= 0.6745 \times \sqrt{\frac{\text{sum of squares of apparent errors.}}{n(n-1).}}$$

SCIENTIFIC USE OF INSTRUMENTS.

Here n denotes the number of observations; and 0·6745 is the modulus adopted.

The method may be illustrated by the following example of the probable error of the result of comparing the bronze Standard yard No. 6 with the Imperial Standard (see Appendix to Fifth Report of Standards Commission, p. 103) when the mean result of nine comparisons had been found to be No. 6 = IS − 1·1 division of the micrometer (or 0·000035057 inch, the value of 1 division being 0·00003187 inch:—

Number of Observation.	Result of each Observation.	Difference from Mean, or Apparent Error.	Squares of Apparent Error.
	Div.	Div.	
1	± 0	1·1	1·21
2	− 2	0·9	0·81
3	− 1	0·1	0·01
4	− 1	0·1	0·01
5	± 0	1·1	1·21
6	− 1	0·1	0·01
7	− 1	0·1	0·01
8	− 1	0·1	0·01
9	− 3	1·9	3·61
Mean =	− 1·1	... Sum =	6·89

$$\frac{\text{Sum of squares}}{n - 1} = \frac{6 \cdot 89}{8} = 0 \cdot 8612$$

Square root of 0·8612 = 0·928

∴ Probable error of result of a single observation
= 0·6745 × 0·928 = ± 0·6259 div., or ± 0·00001995 in.

(IV.) Square root of n observations = $\sqrt{9} = 3$.

∴ Probable error of mean result

$$= \frac{0 \cdot 6259}{3} = \pm 0 \cdot 209 \text{ div., or } \pm 0 \cdot 00000665 \text{ in.}$$

The computation is much facilitated by using Barlow's Table of Squares, Square Roots, &c.

Conclusion.

114. In concluding this discussion of so large a subject as the science of weighing and measuring and the standards of weight and measure, it remains only to add that the great extent and variety of the different matters comprised under these heads, and appearing to deserve notice, have rendered it indispensable to treat many of them briefly and imperfectly, and that much has been omitted in order to bring the volume within the limits prescribed. The object has been to give as much instructive information as the limited space would allow in relation to the standards of weights and measures in use at different periods in various countries, and more particularly to call attention to the scientific basis of our existing standards of weight and measure ; and also to describe the construction of instruments of precision required for the accurate comparison of standards, and to explain the theory and practice of scientific weighing and measuring.

www.ingramcontent.com/pod-product-compliance
Lightning Source LLC
Chambersburg PA
CBHW020920230426
43666CB00008B/1515